Ernst Schering Foundation Symposium
Proceedings 2007-3
Sparking Signals

Ernst Schering Foundation Symposium
Proceedings 2007-3

Sparking Signals

Kinases as Molecular Signaltransducers and Pharmacological Drug Targets in Inflammation

G. Baier, B. Schraven, U. Zügel, A. von Bonin
Editors

With 34 Figures

Series Editors: G. Stock and M. Lessl

Library of Congress Control Number: 2007943076

ISSN 0947-6075

ISBN 978-3-540-73500-7 Springer Berlin Heidelberg New York

Springer is a part of Springer Science+Business Media
springer.com

Cover design: design & production, Heidelberg
Typesetting and production: LE-TEX Jelonek, Schmidt & Vöckler GbR, Leipzig
21/3180/YL – 5 4 3 2 1 0 Printed on acid-free paper

Preface

Almost all aspects of cell life and action are controlled by the reversible phosphorylation of proteins. The human genome projects have revealed that about one-third of mammalian proteins contain covalently bound phosphate, and there are likely to be around 570 protein kinases encoded by the human genome. Protein kinases, classified as either protein tyrosine kinases or serine/threonine kinases, have crucial roles in immune cell signaling, including T cell and Toll-like receptor signaling, that represent central pathways for adoptive and innate immunity. In these immunoreceptor-related pathways, kinases interact sequentially with substrate proteins, which by phosphorylation become activated to allow effective signal transduction to further downstream targets that are directly or indirectly involved in controlling proliferation, homing, and survival of immune cells. Protein kinases therefore play a pivotal role in the initiation, propagation, and regulation of immunological responses. The impressive progress in the past few years in the field of kinases has led to an improved understanding of the role and function of protein kinases as powerful signal transducers for the regulation of immune cell effector functions; kinases are now understood as key mediators for the induction of pathogenesis in a variety of immunological diseases.

The rapid advances in our understanding of the molecular structures of kinases, as well as the advances to unveil the biochemical tuning of kinases, are also the basis for the development of innovative therapeutic agents that target defined protein kinases involved in inflammatory diseases and cancer. The design of kinase inhibitors focuses mainly on the ATP-binding pocket and less conserved surrounding pockets and exploits differences in kinase structure to achieve selectivity. Almost all currently known small inhibitor molecules of protein kinases bind to the active ATP-binding site and act as competitive inhibitors. The high degree of structural similarity in the ATP-binding pocket of protein kinases, however, presents a major challenge to the development of selective inhibitors. Nevertheless, targeting specific kinases that are overexpressed or overactive in disease should allow for selective treatment as shown, e.g., with the success of the specific Abelson murine leukemia viral oncogene kinase inhibitor Gleevec in the treatment of chronic myelogenous leukemia.

The aim of this workshop was to present cutting-edge science in the fast developing kinase field, which is of special interest to basic researchers and the pharmaceutical industry. Within the rather broad field of kinase research, the focus was laid on Toll-like receptor and T cell receptor-mediated signaling. Luckily we were able to bring together internationally renowned experts from both academia and industry, who openly presented and discussed their most recent findings. Most of the presenting scientists agreed to publish their contributions in this book. We are convinced that the scientific forum made available during this kinase workshop allows us a better understanding of signaling events in relevant immune cell subsets, backed up as they are by innovative molecular, structural, and technical, e.g., high-throughput screening, aspects discussed during the meeting. Workshops such as this kinase symposium in Potsdam, therefore, should ultimately have a substantial impact on drug discovery, in particular for developing novel classes of selective inhibitors for the treatment of inflammatory diseases and cancer.

Finally, we would like to thank the Ernst Schering Foundation for the professional support in preparing the meeting, as well as during the workshop, which allowed the open and positive atmosphere for discussion that was present throughout the whole conference in a beautiful, perfectly fitting location close to Potsdam.

Gottfried Baier
Burkhart Schraven
Ulrich Zügel
Arne von Bonin

Contents

List of Editors and Contributors

Editors

Baier, G.
Department for Medical Gentics & Molecular
and Clinical Pharmacology, Medical University of Innsbruck,
Schoepfstr. 41, 6020 Innsbruck, Austria
(e-mail: Gottfried.Baier@i-med.ac.at)

Schraven, B.
Institute of Molecular and Clinical Immunology,
Otto-von-Guericke University,
Leipziger Strasse 44, 39120 Magdeburg, Germany
(e-mail: Burkhart.Schraven@med.ovgu.de)

von Bonin, A.
CMR, Bayer Schering Pharma AG, 13342 Berlin, Germany
(e-mail: arnevon.bonin@bayerhealthcare.com)

Zügel, U.
Common Mechanism Research Early Projects,
Bayer Schering Pharma AG, 13342 Berlin, Germany
(e-mail: Ulrich.Zuegel@bayerhealthcare.com)

Contributors

Bantscheff, M.
Cellzome AG, Meyerhofstrasse 1, 69117 Heidelberg, Germany

de Diego, J.L.
Department for Cellular Microbiology,
Max-Planck Institute for Infection Biology, 10117 Berlin, Germany
(e-mail: diego@mpiib-berlin.mpg.de)

Drewes, G.
Cellzome AG, Meyerhofstrasse 1, 69117 Heidelberg, Germany
(e-mail: gerard.drewes@cellzome.com)

Gerold, G.
Department for Cellular Microbiology,
Max-Planck Institute for Infection Biology, 10117 Berlin, Germany
(e-mail: gerold@mpiib-berlin.mpg.de)

Glück, A.
Novartis Institute for BioMedical Research,
Postfach, 4002 Basel, Switzerland

Gram, H.
Novartis Institute for BioMedical Research,
Postfach, 4002 Basel, Switzerland

Gunzer, M.
Institute of Molecular and Clinical Immunology,
Otto-von-Guericke University,
Leipziger Strasse 44, 39120 Magdeburg, Germany
(e-mail: matthias.gunzer@med.ovgu.de)

Hopf, C.
Cellzome AG, Meyerhofstrasse 1, 69117 Heidelberg, Germany

Jhoti, H.
Astex Therapeutics, 436 Science Park, Milton Road,
Cambridge CB 0QA, UK
(e-mail: h.jhoti@astex-therapeutics.com)

Joyce, C.
Novartis Institute for BioMedical Research,
Postfach, 4002 Basel, Switzerland

Kinzel, B.
Novartis Institute for BioMedical Research,
Postfach, 4002 Basel, Switzerland

Koretzky, G.A.
Department of Pathology and Laboratory Medicine,
Division of Rheumatology, Abramson Family Cancer
Research Institute, University of Pennsylvania, 415 BRBII/III,
421 Curie Blvd., Philadelphia, PA, 19194, USA
(e-mail: Koretzky@mail.med.upenn.edu)

Koziczak-Holbro, M.
Novartis Institute for BioMedical Research,
Postfach, 4002 Basel, Switzerland
(e-mail: magdalena.koziczak@novartis.com)

Kruse, U.
Cellzome AG, Meyerhofstrasse 1, 69117 Heidelberg, Germany

Lindquist, J.A.
Institute of Molecular and Clinical Immunology,
Otto-von-Gucricke University,
Leipziger Strasse 44, 39120 Magdeburg, Germany
(e-mail: jon.lindquist@med.ovgu.de)

Müller, M.
Novartis Institute for BioMedical Research,
Postfach, 4002 Basel, Switzerland

Olenchock, B.A.
Department of Pathology and Laboratory Medicine,
Division of Rheumatology, Abramson Family Cancer
Research Institute, University of Pennsylvania, 415 BRBII/III,
421 Curie Blvd., Philadelphia, PA, 19194, USA

Vestweber, D.
Max-Planck-Institute of Molecular Biomedicine,
Röntgenstr. 20, 48149 Münster, Germany
(e-mail: vestweb@mpi-muenster.mpg.de)

Zhong, X.P.
Department of Pathology and Laboratory Medicine,
Division of Rheumatology, Abramson Family Cancer
Research Institute, University of Pennsylvania, 415 BRBII/III,
421 Curie Blvd., Philadelphia, PA, 19194, USA

Zychlinsky, A.
Department for Cellular Microbiology,
Max-Planck Institute for Infection Biology, 10117 Berlin, Germany
(e-mail: zychlinsky@mpiib-berlin.mpg.de)

Ernst Schering Foundation Symposium Proceedings, Vol. 3, pp. 1–28
DOI 10.1007/2789_2007_060

Published Online: 18 December 2007

Proteomics-Based Strategies in Kinase Drug Discovery

M. Bantscheff, C. Hopf, U. Kruse, G. Drewes(✉)

Cellzome AG, Meyerhofstrasse 1, 69117 Heidelberg, Germany
email: *gerard.drewes@cellzome.com*

Abstract. Studies of drug action classically assess biochemical activity in settings which typically contain the isolated target only. Recent technical advances in mass spectrometry-based analysis of proteins have enabled the quantitative analysis of sub-proteomes and entire proteomes, thus initiating a departure from the traditional single gene—single protein—single target paradigm. Here, we review chemical proteomics-based experimental strategies in kinase drug discovery to analyse quantitatively the interaction of small molecule compounds or drugs with a defined sub-proteome containing hundreds of protein kinases and related proteins. One novel approach is based on 'Kinobeads'—an affinity resin comprised of a cocktail of immobilized broad spectrum kinase inhibitors—to monitor quantitatively the differential binding of kinases and related nucleotide-binding proteins in the presence and absence of varying concentrations of a lead compound or drug of interest. Differential binding is detected by high throughput and sensitive mass spectroscopy techniques utilizing isobaric tagging reagents (iTRAQ), yielding quantitative and detailed target binding profiles. The method can be applied to the screening of compound libraries and to selectivity profiling of lead compounds directly against their endogenously expressed targets in a range of cell types and tissue lysates. In addition, the method can be used to map drug-induced changes in the phosphorylation state of the captured sub-proteome, enabling the analysis of signalling pathways downstream of target kinases.

1 Introduction

Traditionally, biochemical assays of drug action assess activity on the isolated, purified target. Typically, recombinant enzymes or protein fragments are used instead of the full-length endogenous protein. To correlate the activity of a compound determined in such assays with pharmacodynamic efficacy is often not straightforward, as an isolated recombinant protein fragment may not accurately reflect the native conformation and activity of the target in its physiological context (Hall 2006). Because of the lack of interacting regulatory proteins, expression of alternative splice forms, or incorrect folding or post-translational modifications, results from *in vitro* experiments are not always predictive for the effects of a compound or drug in cell-based or animal model systems. Moreover, although lead compounds are traditionally optimized against a single protein, many drugs act on multiple targets (Morphy et al. 2004). Such 'off-targets' may produce toxic side effects, but might also increase the therapeutic potential of a drug. Multiple targets and off-targets are likely to be the rule—rather than the exception—in large target classes with a high degree of structural conservation around the 'druggable' binding site, such as kinases, enzymes which catalyse the transfer of phosphate groups from ATP to the hydroxyl groups of proteins, lipids, or sugars. Protein kinases represent a class of drug targets of increasing importance particularly in oncology and inflammation (Cohen 2002). Small molecule inhibitors directed at the ATP-binding site of kinases are not likely to be selective for a single kinase, because there are around 500 protein kinases and more than 2000 other ATP- and purine-binding proteins encoded in the human genome, many of which share similarly shaped ATP binding pockets (Manning et al. 2002; Haystead 2006). Conventional strategies by and large rely on assay panels of 20–100 recombinant enzymes, and in addition on cell-based model systems, to address compound potency, selectivity and potential off-target liabilities, rather than to determine the *bona fide* targets of a drug in a direct, more unbiased manner (Fabian et al. 2005; Fliri et al. 2005).

Recent progress in affinity-based proteomic strategies has enabled the direct determination of protein-binding profiles of small molecule drugs under more physiological conditions (Szardenings et al. 2004). To date, most methods rely on the attachment of labels to the compound (immobilization, fluorescent or affinity tags) or to the proteins (Fabian et al. 2005; Godl et al. 2003; Knockaert et al. 2000), which may cause artefacts because of the altered properties of the modified compound or the labelled protein. We here describe how technical and methodical advances enabled a strategy involving the capturing of a defined sub-proteome, consisting of a large fraction of the expressed kinome and related nucleotide-binding proteins, on a mixed inhibitor matrix (*kino*me *beads* or Kinobeads), and subsequent quantitative analysis of this defined sub-proteome by mass spectrometry (Bantscheff et al. 2007). This methodology allows the parallel determination of protein affinity profiles of small molecule inhibitors in any cell type or primary tissue as well as the differential mapping of drug-induced changes of phosphorylation events on the captured sub-proteome.

2 Quantitative Mass Spectrometry-Based Proteomics

Proteomics is "the analysis of complete complements of proteins. Proteomics not only includes the identification and quantification of proteins, but also the determination of their localization, modifications, interactions, activities, and, ultimately, their function" (Fields 2001). Originally, proteomics approaches aimed at the identification of a list of typically several hundred proteins expressed in a given tissue or body fluid at a given time under a defined set of conditions. However, the analysis of complex proteomes remains daunting (Huber 2003), as body

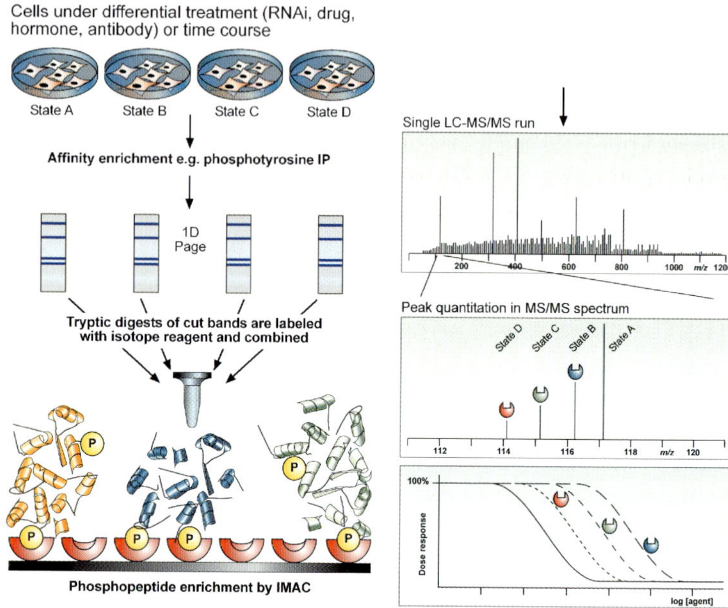

Fig. 1. Multiplexed differential analysis of sub-proteomes by quantitative mass-spectrometry. Up to four different treatments (e.g. hormones, siRNA probes, different compounds, several concentrations of one compound) are applied to a cell line in culture. After enrichment of proteins of interest (e.g. phosphotyrosine immunoprecipitation) proteins are separated by standard one-dimensional SDS polyacrylamide gel electrophoresis and the gel is stained, for example, by colloidal Coomassie dye. Depending on the type of experiment, individual protein bands are cut from the gel, or alternatively, the entire gel is cut into slices across the full separation range. Gel slices are then digested with trypsin and subsequently, the resulting peptide mixtures are labelled using stable isotope-containing tagging reagents (e.g. iTRAQ). After labelling, samples from the differentially treated conditions are mixed and subjected to enrichment of phosphopeptides. These peptides are then analysed using nano-LC separation and tandem mass spectrometry. Tandem mass spectra are used for identification of phosphorylated peptides and quantification is inferred from relative intensities of the low mass signature ions present in the MS/MS spectra

fluids or tissue lysates may contain in excess of 100,000 different proteins, with their relative abundances covering more than seven orders of magnitude, and the general use of 'complete proteome' approaches in drug discovery assays remains challenging (Rappsilber et al. 2002). Established methods tend to be biased towards proteins expressed at high levels, which are mostly structural and house-keeping proteins, and to discriminate against the typical 'signalling' proteins which are often expressed at low levels, with an additional bias against membrane proteins (Rappsilber et al. 2002). In contrast, the analysis of sub-proteomes, biochemically enriched samples of reduced complexity (tens to hundreds of proteins) that share functional context, has made significant progress recently and is being pursued more and more in quantitative fashion (Ong and Mann 2005). This chapter focuses on affinity proteomics methods that utilize affinity chromatography-based strategies for the analysis of sub-proteomes. The combination of novel affinity-based techniques with liquid chromatography (LC)-coupled mass spectrometry has become increasingly successful in analysing protein–protein interactions and protein complexes, in analysing post-translational modifications (e.g. 'phosphoproteomics'), and in elucidating the interaction of therapeutic compounds with their targets ('chemical proteomics'). Recent advances in the ability of mass spectrometry to quantify precisely proteins from complex samples now enables quantitative studies on the regulation of proteins and protein complexes under differential conditions such as hormone or drug treatment (Ong and Mann 2005; Zieske 2006). At present, stable isotope labelling techniques represent the gold standard in quantitative proteomics. There are several different methods that share the underlying principle that a sample of interest is labelled with heavy isotope containing tags or amino acids, and mixed with one or more reference samples that remain either unlabeled or tagged with a lighter isotope (Bantscheff et al. 2004; Gerber et al. 2003; Gygi et al. 1999; Ong et al. 2002; Ross et al. 2004; Schmidt et al. 2005; Zieske 2006). After tryptic digestion, mass spectrometric analysis is performed and proteins are quantified by comparing the ion intensities of heavy versus light forms of isotopically labelled peptides. The iTRAQ method uses a set of four amine-reactive isobaric tagging reagents, and quantification is based on low mass signature ions in tandem mass spectra. The multiplexing capability of the iTRAQ reagents is particularly useful for a variety of applications including temporal analysis or the elucidation of concentration dependency of drug-induced-protein expression, modifications, and interactions, and for the discovery and elucidation of disease markers or markers for drug efficacy. A representative experimental strategy is illustrated in Fig. 1.

3 Chemical Proteomics: Systematic Analysis of Protein–Compound Interactions

Here, we use the term chemical proteomics for methodologies that measure the interaction of small molecule compounds with the proteome, or with defined sub-proteomes (Ding et al. 2003; Szardenings et al. 2004). These approaches, often based on classical affinity labelling or affinity chromatography procedures, are typically applied in drug discovery to elucidate the efficacy targets of compounds with an unknown mechanism of action. For some interesting examples, see the classical work by Schreiber and colleagues on the mechanism of immunosuppressants (Brown et al. 1994; Harding et al. 1989) or the more recent chemogenomic approach to the Wnt pathway (Emami et al. 2004). At early stages in the drug discovery process, the target of hit compounds originating from phenotypic screens in cell-based assays or whole animal studies in most cases is unknown, which makes it much more difficult for the medicinal chemist to optimize such hit compounds (Burdine and Kodadek 2004). However, even lead compounds and approved drugs sometimes lack an established mode of action. Also, it is important to note that small molecule drugs are rarely specific for a single target; they may bind to and modulate other proteins with similarly shaped binding pockets, and they will usually bind to members of the body's complement of xenobiotics—modifying enzymes. Knowledge about these 'off-targets' may alert the researcher to the risk of certain side effects, and may even be used for the reprofiling of the drug for a different therapeutic indication. The general principle of the direct determination of the profile of proteins binding to a given compound of interest is theoretically straightforward (Darvas et al. 2004; Graves et al. 2002). Compounds are tethered to a solid support, covalently or via affinity labels containing biotin, using a suitable chemical linker, and

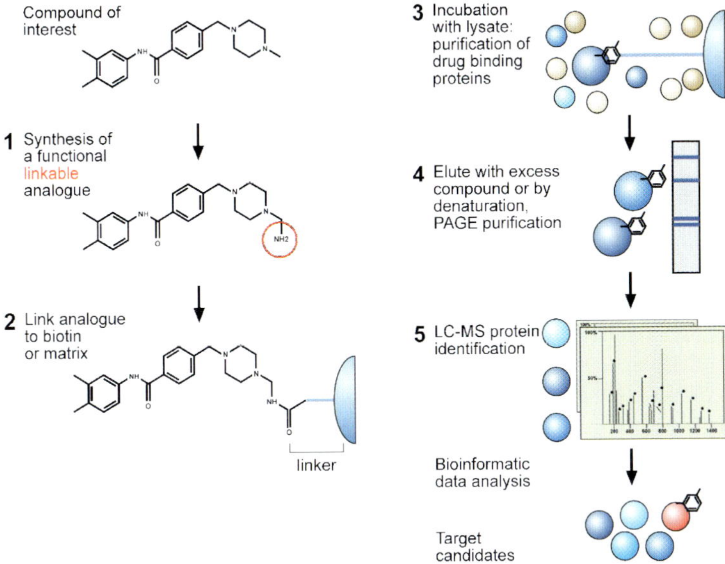

Fig. 2. Drug-affinity purifications. The compounds of interest are tethered to a solid support, covalently or via biotin-avidin methods, using a suitable chemical linker. It is often necessary to synthesize one or more analogues of the original compound, which contain the necessary functional group for coupling, and to test these analogues to ensure that their activity is preserved. The resulting drug-affinity matrix is incubated with cell or tissue lysate to capture specific drug-interacting proteins. After washing, bound proteins are eluted either by a large excess of the free compound or simply by denaturation, and are subsequently identified by LC-coupled tandem mass spectrometry. Usually, several experimental parameters such as the length and nature of the linker, the ligand density of the resulting affinity matrix, and the amount of tissue lysate, will require some empirical optimization in order to achieve an acceptable background level of non-specific proteins. Note that non-specific proteins interacting with the affinity matrix can be determined in a parallel experiment where an excess of the free compound of interest is present during the incubation with the lysate

incubated with cell or tissue lysate to capture specific interacting protein targets. After washing, bound proteins are eluted either by an excess of the free compound or simply by denaturation, and are subsequently identified by mass spectrometry (Fig. 2). Non-specific proteins interacting with the affinity matrix can be determined in a parallel experiment where an excess of the free compound of interest is present during incubation with the lysate. However, a few potential complications must be taken into account. The main types of functional groups that permit the attachment of the linker—amines, carboxylic acids, thiols, and hydroxyls—are often not present in the compound of interest, or cannot be modified without the complete or partial loss of binding to the target. Hence, it is often required to synthesize one or more analogues containing the necessary functional group and to test these analogues to make sure that their activity is preserved (Daub 2005). Usually, several other experimental parameters such as the length and nature of the linker, the ligand density of the resulting affinity matrix, and the amount of tissue lysate, will all require optimization in order to achieve an acceptable background level of non-specific proteins. Target candidates identified by affinity labelling and affinity chromatography can fall into one of many categories, including efficacy targets, mediators of adverse events, proteins involved in absorption, distribution, metabolism, excretion of the compound, or proteins that do not directly interact with the compound but with any of the aforementioned proteins. Complexity of the experimental outcome makes detailed bioinformatics analysis and functional confirmation (using, for instance, RNA interference in a cell-based assay) often mandatory. In reality, affinity purification strategies do not work equally well for proteins from different target classes. The approach appears to be particularly suited to the profiling of protein and lipid kinases, ATP-binding proteins (e.g. chaperones, ATPases, helicases, or transporters), flavoenzymes, phosphodiesterases, proteases, and peptidomimetics and other agents targeting protein–protein interactions. In contrast, the approach appears to be less suited for ion channels, nuclear receptors, or G-protein coupled receptors. Here, we will focus on the application to kinases and related ATP-binding proteins.

In our laboratory, we have to date profiled a collection of more than 100 ATP-competitive kinase inhibitors including chemical scaffolds, research tool compounds, drug candidates in development, as well as approved drugs, by tethering them to a solid support using a suitable chemical linker, and incubation with cell or tissue lysates to capture and identify their interacting proteins (Bantscheff et al. 2007). Some exam-

Fig. 3. Examples of affinity purification data from a number of immobilized kinase inhibitors. The research tool inhibitors Bis (III) indolyl maleimide (a protein kinase C inhibitor), purvalanol B (a cyclin-dependent kinase inhibitor), CZC8004 and staurosporine (both pan-kinase inhibitors); and linkable analogues of the drugs or drug candidates PD173955 (Src kinases), vandetanib (VEGFR, EGFR), sunitinib (VEGFR, PDGFR, Flt3, KIT), Ro 320–1195 (p38 MAP kinase), imatinib (ABL, PDGFR, KIT), gefitinib (EGFR), pelitinib (EGFR), and lapatinib (EGFR, Her-2) were immobilised and mixed with lysates from HeLa or K562 cells. After washing, and elution with detergent, bound proteins were identified by mass spectrometry. The number of spectrum-to-sequence matches was translated into a 'heat map', serving as a semi-quantitative indication of the amount of protein captured

ples are shown in Fig. 3. In cases where the compounds did not contain functional groups suitable for covalent coupling to an affinity matrix while preserving activity, analogues containing primary amino groups were designed and synthesized. After separation of the beads from the lysate, bound proteins were digested with trypsin, and identified by LC-coupled tandem mass spectrometry. In cases where the known targets of the compounds are expressed in the cell or tissue samples under study, these targets were frequently identified, and moreover, novel potential targets were identified. As shown in Fig. 3, some of the compounds interacted rather selectively with few kinase targets while others displayed low apparent selectivity. For instance, the immobilized analogue of the ABL inhibitor imatinib, or the analogues of the epidermal growth factor receptor inhibitors gefitinib and lapatinib, bind few other kinases beyond their known targets. In contrast, several tool compounds, but also some immobilized analogues of drugs, e.g. sunitinib and vandetanib, bind to a much larger number of kinases. While the qualitative affinity profiles of immobilized compounds reveal novel target candidates, they are less suitable for the validation of inhibitor specificity for a number of reasons. First, the results obtained for the immobilized molecule may not relate directly to the original compound owing to altered potency and selectivity due to the attachment of the linker. Moreover, the resulting binding profiles are biased towards abundant proteins, which

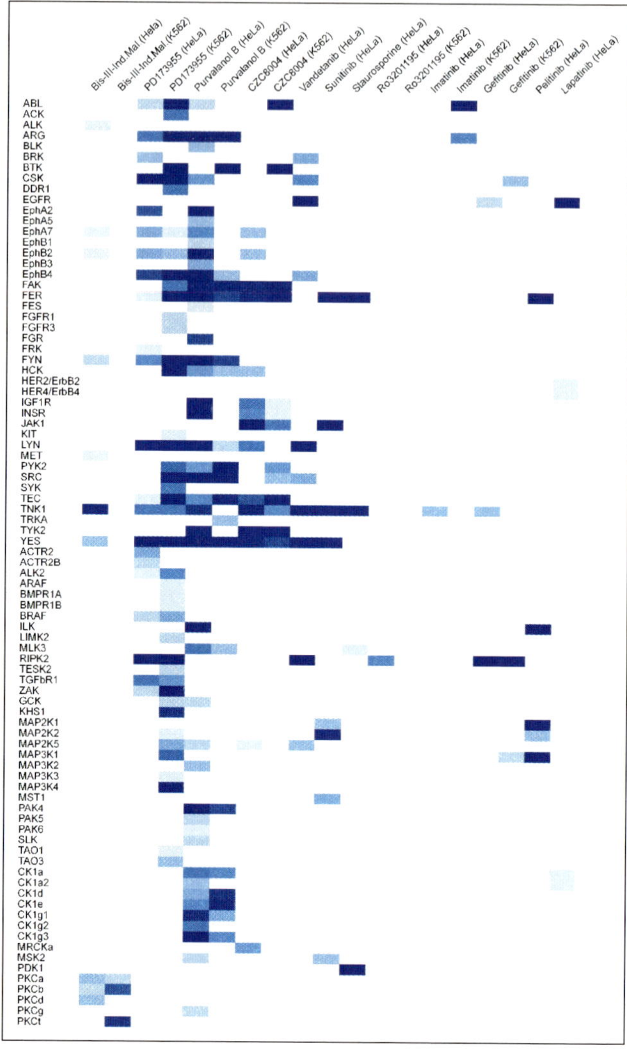

Bis-III-Ind.Mal (Hela)
Bis-III-Ind.Mal (K562)
PD173955 (HeLa)
PD173955 (K562)
Purvalanol B (HeLa)
Purvalanol B (K562)
CZC8004 (HeLa)
CZC8004 (K562)
Vandetanib (HeLa)
Sunitinib (HeLa)
Staurosporine (HeLa)
Ro3201195 (HeLa)
Ro3201195 (K562)
Imatinib (HeLa)
Imatinib (K562)
Gefitinib (HeLa)
Gefitinib (K562)
Pelitinib (HeLa)
Lapatinib (HeLa)
ABL
ACK
ALK
ARG
BLK
BRK
BTK
CSK
DDR1
EGFR
EphA2
EphA5
EphA7
EphB1
EphB2
EphB3
EphB4
FAK
FER
FES
FGFR1
FGFR3
FGR
FRK
FYN
HCK
HER2/ErbB2
HER4/ErbB4
IGF1R
INSR
JAK1
KIT
LYN
MET
PYK2
SRC
SYK
TEC
TNK1
TRKA
TYK2
YES
ACTR2
ACTR2B
ALK2
ARAF
BMPR1A
BMPR1B
BRAF
ILK
LIMK2
MLK3
RIPK2
TESK2
TGFbR1
ZAK
GCK
KHS1
MAP2K1
MAP2K2
MAP2K5
MAP3K1
MAP3K2
MAP3K3
MAP3K4
MST1
PAK4
PAK5
PAK6
SLK
TAO1
TAO3
CK1a
CK1a2
CK1d
CK1e
CK1g1
CK1g2
CK1g3
MRCKa
MSK2
PDK1
PKCa
PKCb
PKCd
PKCg
PKCt

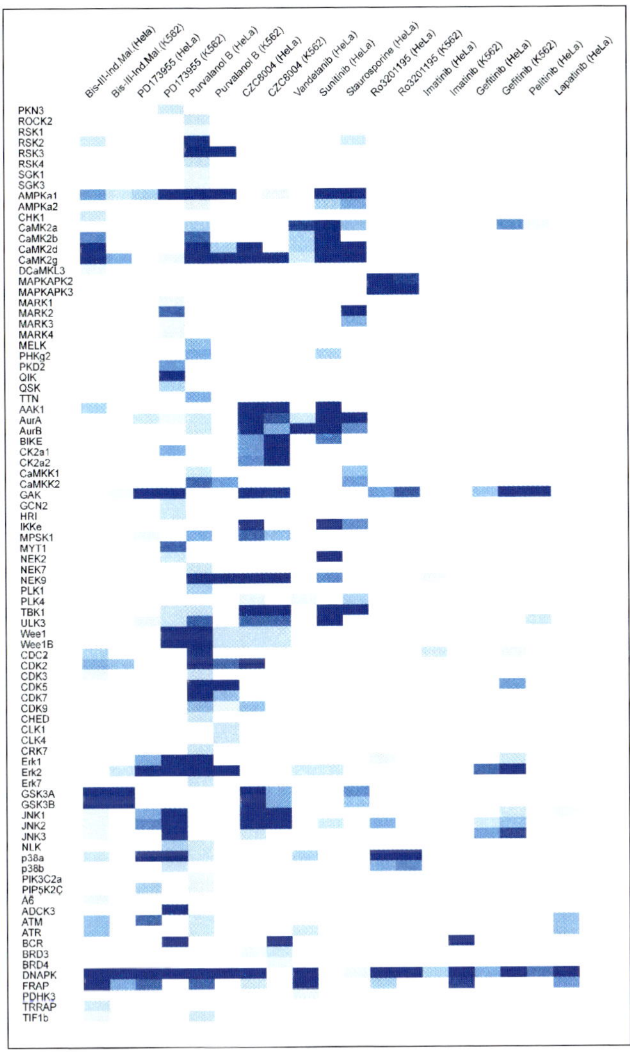

Fig. 3. (continued)

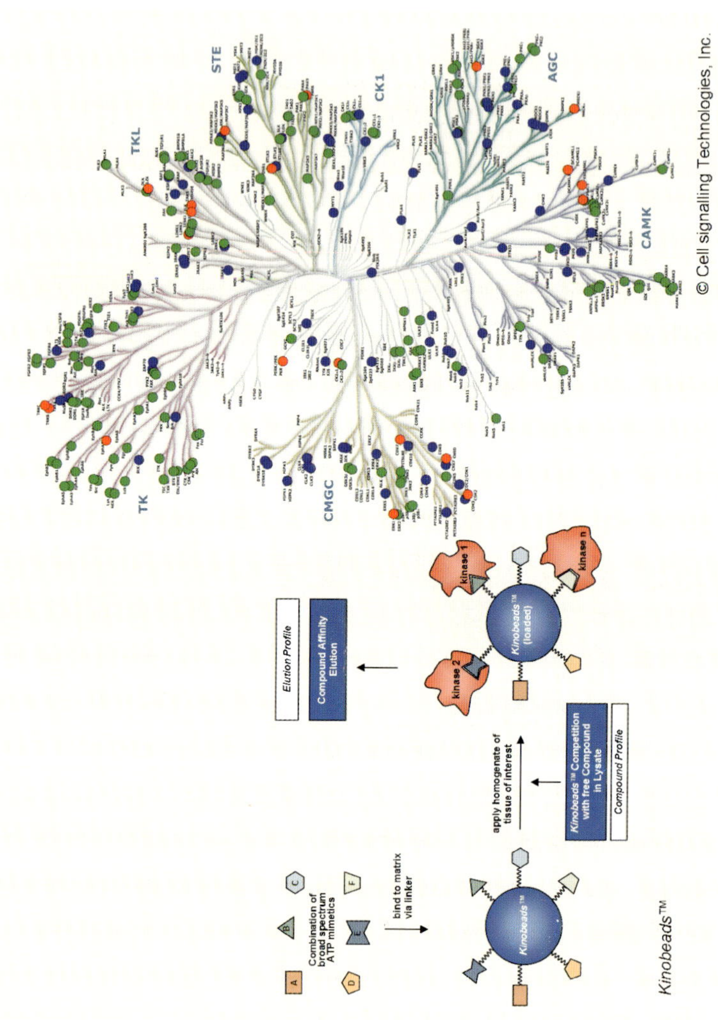
A
B
C
D
E
F
Combination of broad spectrum ATP mimetics
bind to matrix via linker
Kinobeads™
apply homogenate of tissue of interest
Kinobeads™ Competition with free Compound in Lysate
Compound Profile
kinase 1
kinase 2
kinase n
Kinobeads™ (loaded)
Compound Affinity Elution
Elution Profile
Kinobeads™
TK
TKL
STE
CK1
AGC
CAMK
CMGC
© Cell signalling Technologies, Inc.

are frequently only weakly affected in subsequent activity-based assays (Godl et al. 2003; Knockaert et al. 2000; Wissing et al. 2004).

4 Kinobeads: Application of Quantitative Chemical Proteomics to Kinases

As outlined above, the affinity profiling of compounds by immobilization of suitable analogues is a powerful tool for the identification of target candidates, but it is not ideal for the validation of inhibitor selectivity. Therefore, a novel methodology was developed, which uses immobilized broad-specificity inhibitors as kinase-capturing tools to analyse the interaction of competing 'free' compound with their protein targets in solution (Bantscheff et al. 2007). The method is based on measuring the degree of competition between the unmodified test compound and the immobilized ligands for ATP-binding and related sites on proteins. For an unbiased target profile, a capturing ligand binding to all members of a target class of interest would be required. A previously described method used ATP immobilized via its gamma-phosphate group as affinity ligand (Graves et al. 2002). However, in our experience this approach resulted in the capture of only a small number of kinases

Fig. 4. Principle of the Kinobeads technology. Broad-spectrum ATP-competitive kinase inhibitors are mixed, and immobilized covalently on a bead-based matrix. The matrix is mixed with a tissue lysate and kinases and related proteins are affinity-captured. Any compound of interest can now be profiled against the captured proteins by either including it in the lysate prior to the Kinobeads binding step, or it can be used for affinity-elution of protein targets from the loaded beads. Kinase binding data from the mass spectrometric analysis of Kinobeads purifications from 14 human and rodent cell lines and tissues (human HEK 293, HeLa, Jurkat, K562, Ramos, THP-1, kidney, placenta; mouse heart, liver, brain, muscle, kidney; and rat RBL-2H3) led to the identification of 307 kinases (269 human and 196 murine) across all branches of the phylogenetic tree. Kinases that were found both in human and mouse samples are shown as green dots, while the ones specific for either human or mouse are shown in blue or red respectively). The kinase phylogenetic tree was adapted with permission from Cell Signaling Inc. (www.cellsignal.com)

(<10) and instead was dominated by the binding of heat shock proteins. In our alternative approach, a set of tool compounds that displayed little selectivity and interacted with protein kinases located on different

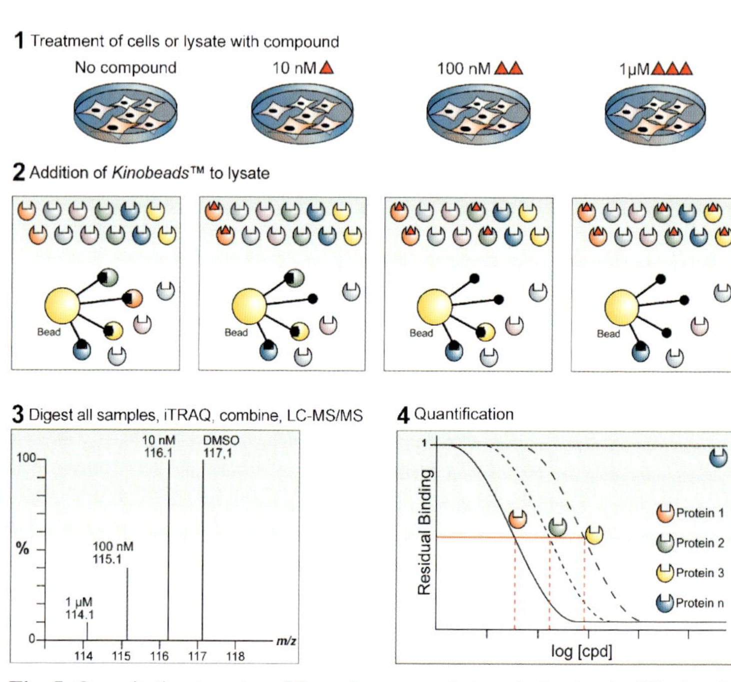

Fig. 5. Quantitative target profiling of compounds in solution by the Kinobeads assay. Cell lysates or cells in culture are treated with vehicle and with compound over a range of concentrations (*upper panel*). Subsequently, proteins are captured on Kinobeads. The 'free' inhibitor competes with the immobilized ligands for ATP-binding or related ligand-binding sites of its targets (*middle panels*). Bound proteins are digested with trypsin and each peptide pool is labelled with iTRAQ reagent (not shown). All four samples are combined and analysed by mass spectrometry. Each peptide gives rise to four characteristic iTRAQ reporter signals indicative of the inhibitor concentration used (*bottom left panel*). For each peptide detected, the decrease of signal intensity compared to the vehicle control reflects competition by the 'free' compound for its target (*bottom right panel*)

branches of the phylogenetic tree was immobilized to create mixed inhibitor resins of up to seven different capturing ligands (Bantscheff et al. 2007). These mixed kinase inhibitor beads (termed kinome beads or Kinobeads) specifically captured a large portion of the expressed kinome. A defined combination of seven broad-spectrum compounds captured a total of 173 and 179 protein kinases from HeLa and K562 cells respectively, in single pull-down experiments analysed using LC/MS-MS analysis. By expanding this approach by using the same matrix with 14 different human and rodent cell lines and tissues, a total of 307 non-redundant protein kinases were identified (Fig. 4). While a slightly lower relative coverage was observed within the serine/threonine kinase branches compared to the tyrosine kinase branch, there were no major gaps. Moreover, Kinobeads do not only capture protein kinases, but bind a defined sub-proteome consisting also of other ATP- and purine-binding proteins such as chaperones, helicases, ATPases, motor proteins, transporters, and metabolic enzymes. However, based on the total number of peptide-to-spectrum matches, it was estimated that kinases account for more than half of the total captured protein amount.

The quantitative profiling by Kinobeads competition binding in solution was applied to three inhibitors of the tyrosine kinase ABL developed for the treatment of chronic myelogenous leukemia; the phase II compound SKI-606 and the marketed drugs imatinib and dasatinib (Weisberg et al. 2007). In one set of experiments, the drugs were added over a range of concentrations from 100 pM to 10 μM to lysates of K562 human erythroleukemia cells, which express the constitutively active BCR-ABL oncogene. Subsequently the treated lysates were sub-

Fig. 6a. Examples of competition binding curves calculated from iTRAQ reporter spectra. Binding of several known and novel targets to Kinobeads is shown as dependent on the addition of imatinib (*blue*), dasatinib (*green*) and SKI-606 (*red*) in K562 cell lysate. Independent quadruplexed experiments (vehicle plus three compound concentrations each) were performed and iTRAQ reporter data were combined to display the dose response over nine different concentrations. The *bar diagram* shows the full Kinase binding profile of the ABL inhibitors gleevec/imatinib (*upper panel*), sprycel/dasatinib (*middle panel*), and SKI-606 in K562 lysate (*lower panel*)

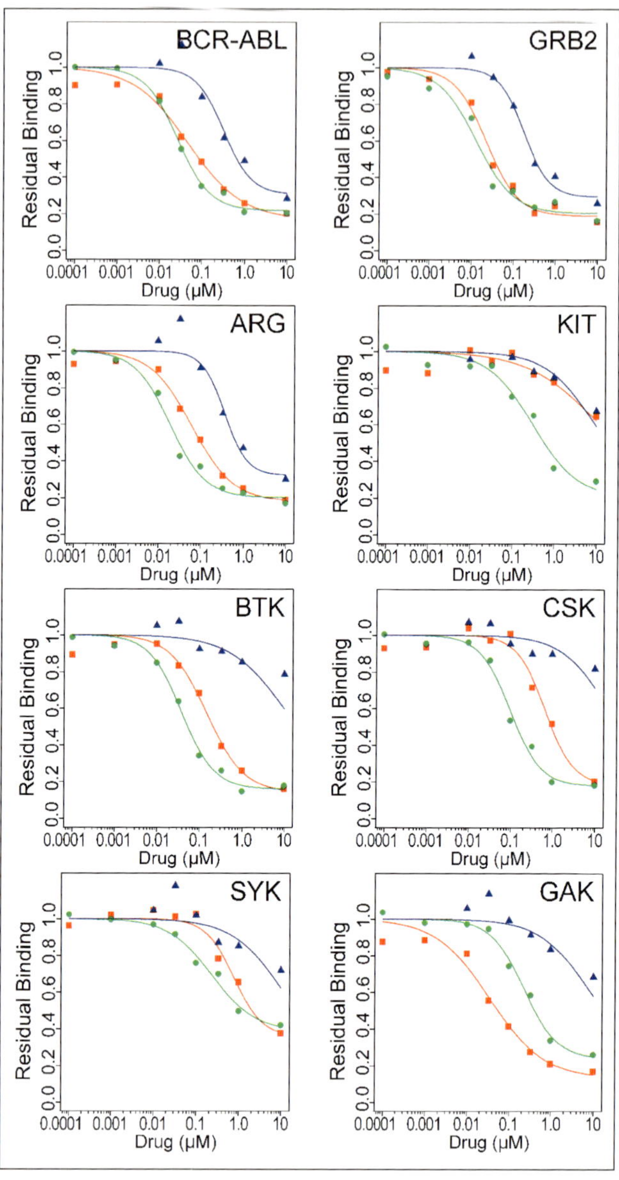
BCR-ABL
GRB2
ARG
KIT
BTK
CSK
SYK
GAK
Residual Binding
0.0
0.2
0.4
0.6
0.8
1.0
0.0001
0.001
0.01
0.1
1.0
10
Drug (μM)

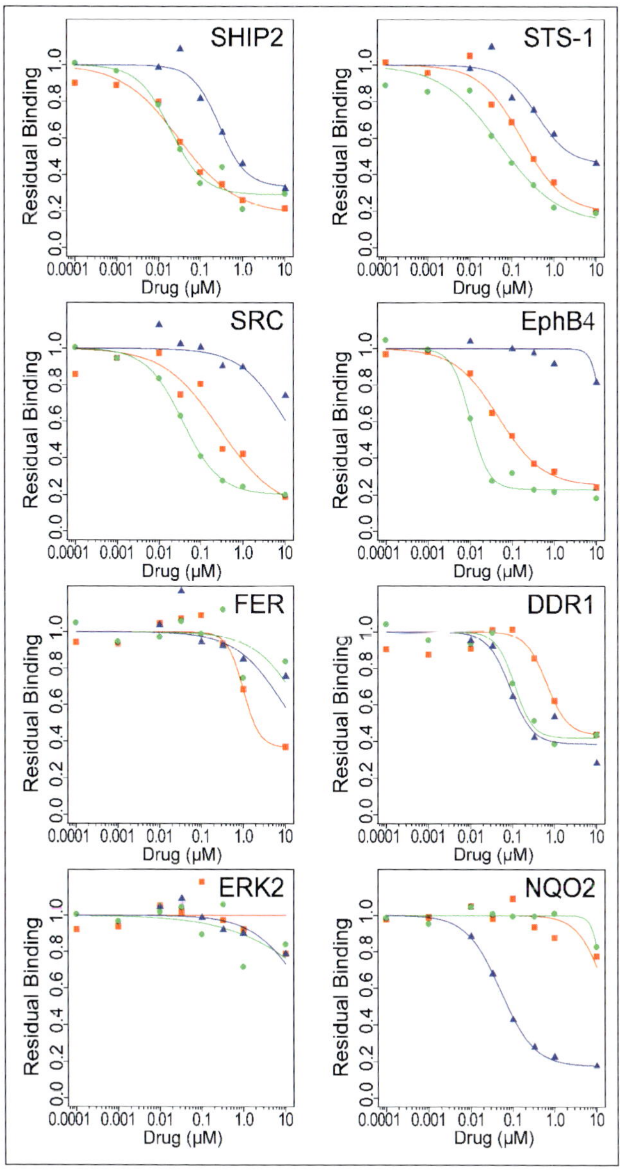
SHIP2
STS-1
SRC
EphB4
FER
DDR1
ERK2
NQO2
Residual Binding
0.0 0.2 0.4 0.6 0.8 1.0
Drug (μM)
0.0001 0.001 0.01 0.1 1.0 10

jected to Kinobeads precipitation. When thc drug in the lysate binds its target and thus blocks the ATP binding site, a reduced amount of the free target is available for capturing on Kinobeads, while the binding of non-targeted kinases and other proteins is unaffected (Fig. 5). The Kinobeads-bound material from each spiking experiment was subjected to tryptic digestion, and peptides were labelled with the four different forms of the iTRAQ reagent (Ross et al. 2004). Subsequently, peptide mixtures were combined and subjected to mass spectrometric quantification by measuring the signal of the iTRAQ reporter ions relative to vehicle-treated lysate. Dose–response binding curves are then computed by analysis of all individual peptide dose–response data obtained for a given protein. The IC_{50} values obtained are largely independent of the affinity of the targets for the immobilized ligands, because the effective concentration of capturing molecules is typically below the range of affinities of the competing compound for its targets (Lowe et al. 1973). Hence data obtained for all proteins in the same samples can be directly compared, which is an advantage compared to IC_{50} values determined in enzyme assays, which depend to a considerable degree on the assay conditions (Knight and Shokat 2005). For the three drugs, dose–response binding profiles for >500 proteins in each sample were generated, including ~150 kinases. For imatinib, 13 proteins exhibited >50% binding reduction on Kinobeads at 1 μM drug in the lysate. Among the competed proteins are ABL/BCR-ABL, ARG, and two novel target candidates, the receptor tyrosine kinase DDR1 (90 nM), and the quinone oxidoreductase NQO2 (43 nM) (Fig. 6). In contrast, dasatinib revealed a broad target profile (46 and 42 proteins respectively showed >50% competition at 1 μM), including the three imatinib targets ABL/BCR-ABL, ARG, and DDR1 (Fig. 7). In addition to kinases, several non-kinase targets were identified, some of which do not contain obvious small molecule binding sites and hence are likely to bind indirectly to the drugs, as proteins which reside in a complex with the drug target are expected to exhibit similar competition behaviour. Indeed this is the case for the BCR-ABL interacting proteins

Fig. 6b. (continued)

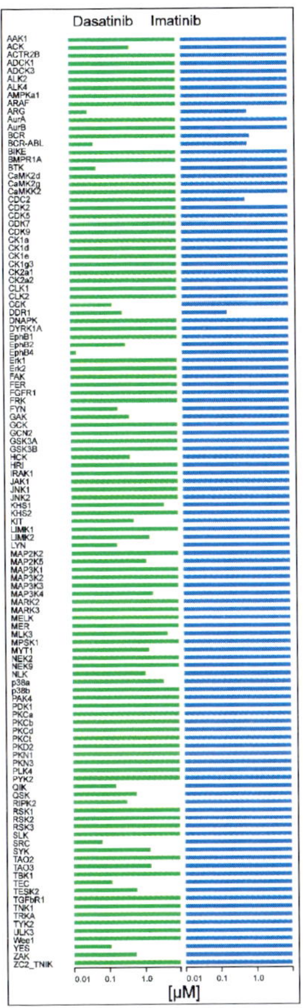

Fig. 7. Kinase binding profiles of the ABL drugs gleevec/imatinib (*right panel*) and sprycel/dasatinib (*left panel*) across a set of protein kinases simultaneously identified from K562 cells. The *bars* indicate the IC_{50} values, defined as the concentration of drug at which half-maximal competition of Kinobeads binding is observed

GRB2, SHC1, and SHIP2. STS-1, an adaptor protein reported to regulate tyrosine kinases by inhibiting the ubiquitin ligase CBL (Kowanetz et al. 2004), also showed a similar dose–response for all three drugs.

Consequently it was proposed as a BCR-ABL/ABL kinase interacting protein.

Several of the newly identified targets were confirmed in enzyme activity assays. More than half of the novel targets were not available in any of the commercial assay panels, but for 10 kinases IC_{50} values in enzyme activity assays were determined and showed a general trend supporting the Kinobeads data. Interestingly, the oxidoreductase NQO2 represents the first non-kinase target of imatinib. The binding of imatinib to NQO2 is specific, as dasatinib and SKI-606 did not compete efficiently. NQO2 is a cytosolic flavoprotein that catalyses the metabolic reduction of quinones and related xenobiotics and protects cells against oxidative stress and neoplasia (Vella et al. 2005). In an enzyme assay using purified NQO2, imatinib displayed potent competitive inhibition ($K_i = 39$ nM). However, because the physiological role of NQO2 is not well understood, it is difficult to predict the consequences of the inhibition of this enzyme.

5 Pathway Profiling: Proteomics-Based Mapping of Post-translational Modifications

The comprehensive analysis of post-translational modifications is expected to uncover another layer of functional complexity underlying cellular systems and has led to a surge in interest in biomarker discovery. Consequently, an arsenal of techniques has been established for the enrichment of post-translationally modified, in most cases phosphorylated, proteins or peptides (Morandell et al. 2006). Following enrichment, these proteins/peptides are further fractionated using gel electrophoresis and/or LC and the modified amino acid residues are identified by tandem mass spectrometry. Because of its technical robustness, immuno-affinity purification of tyrosine phosphorylated proteins or peptides has been used successfully in many fields, for example to identify novel proteins involved in insulin signalling and potentially deregulated proteins in cancer cells, respectively (Rush et al. 2005; Wang et al. 2006). The more general approaches that also allow the identification of serine and threonine phosphorylation sites, immobilized metal affinity chromatography (IMAC) (Moser and White 2006;

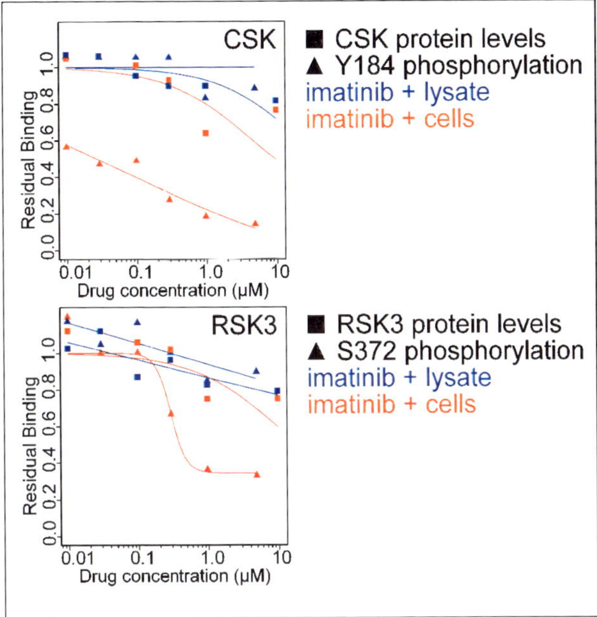

Fig. 8a. Parallel profiling of direct targets and downstream effects of drugs by quantification and phosphorylation analysis of Kinobeads-captured proteins from untreated and drug-treated cells. Dose-dependent reduction of regulatory phosphorylation sites in imatinib-treated K562 cells (*triangles*) or lysates (*squares*) of regulatory sites on Csk (*upper panel*) and RSK2 (*bottom panel*).

Nuhse and Peck 2006) and titanium dioxide affinity chromatography (Larsen et al. 2005; Pinkse et al. 2004) are now the tools of choice for the enrichment of phosphorylated peptides and proteins (Dubrovska and Souchelnytskyi 2005; Kratchmarova et al. 2005; Zhang et al. 2005).

For a better prediction of drug effects it is useful to analyse the impact on the underlying signalling pathways. The mapping of drug-induced post-translational changes in the cellular kinome and its associated proteins can reveal effects of a drug downstream of its direct targets. Potent kinase inhibitors typically exhibit slow off-rates, which permits a powerful variation of the Kinobeads strategy of profiling compounds in a lysate: The compounds are applied over a range of concentrations

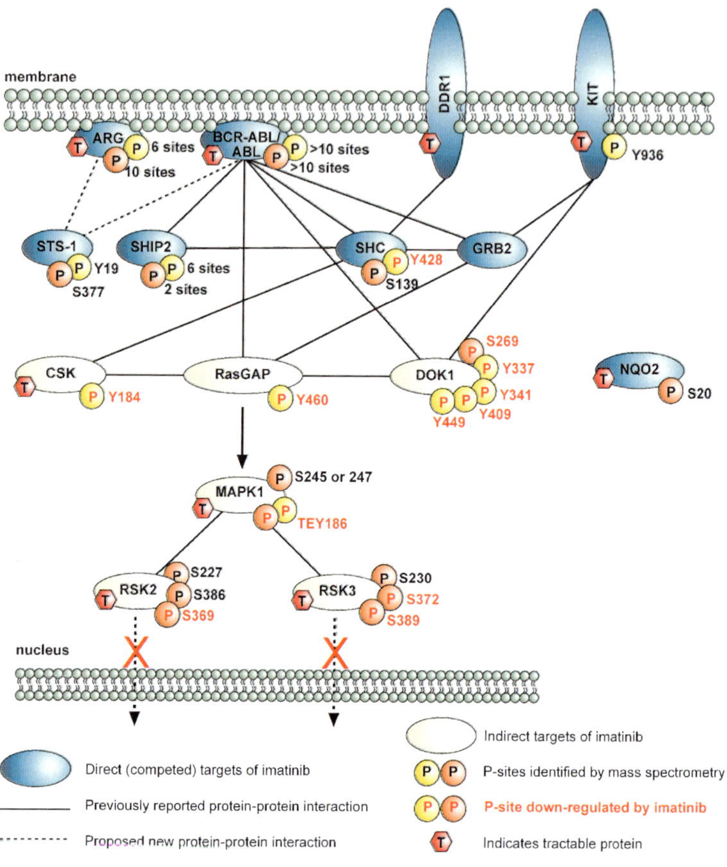

directly to cultured cells for a defined time period, after which cells are lysed and subjected to Kinobeads precipitation. The direct targets of the compound are then identified by the degree of binding competition, analogous to the previous lysate competition experiments. To define not only the direct targets of the drugs but also their downstream effects on signalling pathways, aliquots of the iTRAQ-labelled peptide mixtures from Kinobeads precipitates are applied to IMAC columns to en-

Fig. 8b. Parallel profiling of direct targets and downstream effects of drugs by quantification and phosphorylation analysis of Kinobeads-captured proteins from untreated and drug-treated cells. Proposed mechanism of action of imatinib in K562 chronic myelogenous leukemia cells. Direct targets (*blue symbols*) bind directly to the drug, or are associated in a complex with proteins directly binding and hence exhibit decreased binding to Kinobeads in the presence of the drug. Indirect targets (*white symbols*) represent substrates of the direct targets. They do not bind directly to the drug and hence their binding to Kinobeads is not affected, but they do exhibit reduced phosphorylation of potential or known regulatory sites. Imatinib binds to its direct target, which appears to be a BCR-ABL/GRB2/SHC/SHIP2/STS-1 complex, as all of these proteins are competed by imatinib and dasatinib with similar characteristic potencies. Additional direct imatinib targets are the kinases Arg, DDR1, and KIT, and the oxidoreductase NQO2. Inhibition of the constitutively active BCR-ABL kinase leads to decreased tyrosine phosphorylation of the adaptors SHC and DOK1, and of the GTPase activating protein RasGAP, resulting in down-regulation of the MAP kinase pathway and subsequent prevention of nuclear entry and transcriptional activation of RSK kinases

rich phosphopeptides for subsequent identification and quantification by mass spectrometry. In one example cells were treated with imatinib and 351 tyrosine and serine/threonine phosphorylation sites on 119 different proteins were identified from the Kinobeads-bound material. Only 14 of these sites on nine different proteins exhibited significant down-regulation in response to imatinib (Fig. 8). Indeed, several of these proteins have been implicated in ABL-dependent signalling events, and a large portion of the BCR-ABL signalling pathway (Hantschel and Superti-Furga 2004; Weisberg et al. 2007) is recapitulated in the Kinobeads data. Imatinib binding to its direct target BCR-ABL results not only in the loss of BCR-ABL from Kinobeads, but similarly reduces the amount of the associated signalling proteins GRB2, SHC, and SHIP2. STS-1 is reduced with similar dose–response characteristics and probably represents a novel member of the BCR-ABL signalling complex. The inhibition of ABL kinase activity leads to decreased tyrosine phosphorylation of the adaptors SHC (at Y427) and DOK1 (at several sites), and of the GTPase activating protein RasGAP (at Y460). In turn, MAP kinase is down-regulated (at T184/Y186) leading to reduced phosphorylation of RSK kinases (at S360/377), preventing nuclear translocation and the induction of transcription (Yang et al. 2005).

6 Conclusions

The application of innovative mass spectrometry-based proteomic methodology has transformed two now burgeoning areas in biomedical research: the mapping of protein complexes and the charting of signalling pathways via large-scale analysis of post-translational modifications. The next revolution is anticipated in chemical proteomics, the profiling of small molecule interactions with the proteome. Quantitative chemical proteomic approaches have now been applied to kinase inhibitors, and enable, for the first time, the determination of the binding of small molecules to their targets directly in cells or lysates of relevant tissues. The Kinobeads approach in combination with quantitative mass spectrometry provides a versatile tool to map a drug's direct and indirect targets in a single set of experiments. We anticipate that affinity-based proteomic approaches will prove valuable at various stages of drug discovery as well as in translational studies of drug action in patient tissues. Large-scale proteomic datasets comprising protein complex networks, post-translational modifications, and interactions with pharmacological agents or metabolites, represent an ideal bioinformatic platform for the integration of orthogonal datasets from transcription profiling, and genetics.

References

Bantscheff M, Dumpelfeld B, Kuster B (2004) Femtomol sensitivity post-digest (18)O labeling for relative quantification of differential protein complex composition. Rapid Commun Mass Spectrom 18:869–876

Bantscheff M, Eberhard D, Abraham Y, Bastuck S, Boesche M, Hobson S, Mathieson T, Perrin J, Raida M, Rau C, Reader V, Sweetman G, Bauer A, Bouwmeester T, Hopf C, Kruse U, Neubauer C, Ramsden N, Rick J, Kuster B, Drewes G (2007) A quantitative chemical proteomics approach reveals novel modes of action of clinical ABL kinase inhibitors. Nat Biotechnol 25:1035–1044

Brown EJ, Albers MW, Shin TB, Ichikawa K, Keith CT, Lane WS, Schreiber SL (1994) A mammalian protein targeted by G1-arresting rapamycin-receptor complex. Nature 369:756–758

Burdine L, Kodadek T (2004) Target identification in chemical genetics: the (often) missing link. Chem Biol 11:593–597

Cohen P (2002) Protein kinases–the major drug targets of the twenty-first century? Nat Rev Drug Discov 1:309–315

Darvas F, Dorman G, Krajcsi P, Puskas LG, Kovari Z, Lorincz Z, Urge L (2004) Recent advances in chemical genomics. Curr Med Chem 11:3119–3145

Daub H (2005) Characterisation of kinase-selective inhibitors by chemical proteomics. Biochim Biophys Acta 1754:183–190

Ding S, Wu TY, Brinker A, Peters EC, Hur W, Gray NS, Schultz PG (2003) Synthetic small molecules that control stem cell fate. Proc Natl Acad Sci U S A 100:7632–7637

Dubrovska A, Souchelnytskyi S (2005) Efficient enrichment of intact phosphorylated proteins by modified immobilized metal-affinity chromatography. Proteomics 5:4678–4683

Emami KH, Nguyen C, Ma H, Kim DH, Jeong KW, Eguchi M, Moon RT, Teo JL, Kim HY, Moon SH, Ha JR, Kahn M (2004) A small molecule inhibitor of beta-catenin/CREB-binding protein transcription [corrected]. Proc Natl Acad Sci USA 101:12682–12687

Fabian MA, Biggs WH III, Treiber DK, Atteridge CE, Azimioara MD, Benedetti MG, Carter TA, Ciceri P, Edeen PT, Floyd M, Ford JM, Galvin M, Gerlach JL, Grotzfeld RM, Herrgard S, Insko DE, Insko MA, Lai AG, Lelias JM, Mehta SA, Milanov ZV, Velasco AM, Wodicka LM, Patel HK, Zarrinkar PP, Lockhart DJ (2005) A small molecule-kinase interaction map for clinical kinase inhibitors. Nat Biotechnol 23:329–336

Fields S (2001) Proteomics. Proteomics in genomeland. Science 291:1221–1224

Fliri AF, Loging WT, Thadeio PF, Volkmann RA (2005) Analysis of drug-induced effect patterns to link structure and side effects of medicines. Nat Chem Biol 1:389–397

Gerber SA, Rush J, Stemman O, Kirschner MW, Gygi SP (2003) Absolute quantification of proteins and phosphoproteins from cell lysates by tandem MS. Proc Natl Acad Sci USA 100:6940–6945

Godl K, Wissing J, Kurtenbach A, Habenberger P, Blencke S, Gutbrod H, Salassidis K, Stein-Gerlach M, Missio A, Cotten M, Daub H (2003) An efficient proteomics method to identify the cellular targets of protein kinase inhibitors. Proc Natl Acad Sci USA 100:15434–15439

Graves PR, Kwiek JJ, Fadden P, Ray R, Hardeman K, Coley AM, Foley M, Haystead TA (2002) Discovery of novel targets of quinoline drugs in the human purine binding proteome. Mol Pharmacol 62:1364–1372

Gygi SP, Rist B, Gerber SA, Turecek F, Gelb MH, Aebersold R (1999) Quantitative analysis of complex protein mixtures using isotope-coded affinity tags. Nat Biotechnol 17:994–999

Hall SE (2006) Chemoproteomics-driven drug discovery: addressing high attrition rates. Drug Discov Today 11:495–502

Hantschel O, Superti-Furga G (2004) Regulation of the c-Abl and Bcr-Abl tyrosine kinases. Nat Rev Mol Cell Biol 5:33–44

Harding MW, Galat A, Uehling DE, Schreiber SL (1989) A receptor for the immunosuppressant FK506 is a cis-trans peptidyl-prolyl isomerase. Nature 341:758–760

Haystead TA (2006) The purinome, a complex mix of drug and toxicity targets. Curr Top Med Chem 6:1117–1127

Huber LA (2003) Is proteomics heading in the wrong direction? Nat Rev Mol Cell Biol 4:74–80

Knight ZA, Shokat KM (2005) Features of selective kinase inhibitors. Chem Biol 12:621–637

Knockaert M, Gray N, Damiens E, Chang YT, Grellier P, Grant K, Fergusson D, Mottram J, Soete M, Dubremetz JF, Le RK, Doerig C, Schultz P, Meijer L (2000) Intracellular targets of cyclin-dependent kinase inhibitors: identification by affinity chromatography using immobilised inhibitors. Chem Biol 7:411–422

Kowanetz K, Crosetto N, Haglund K, Schmidt MH, Heldin CH, Dikic I (2004) Suppressors of T-cell receptor signaling Sts-1 and Sts-2 bind to Cbl and inhibit endocytosis of receptor tyrosine kinases. J Biol Chem 279:32786–32795

Kratchmarova I, Blagoev B, Haack-Sorensen M, Kassem M, Mann M (2005) Mechanism of divergent growth factor effects in mesenchymal stem cell differentiation. Science 308:1472–1477

Larsen MR, Thingholm TE, Jensen ON, Roepstorff P, Jorgensen TJ (2005) Highly selective enrichment of phosphorylated peptides from peptide mixtures using titanium dioxide microcolumns. Mol Cell Proteomics 4:873–886

Lowe CR, Harvey MJ, Craven DB, Dean PD (1973) Some parameters relevant to affinity chromatography on immobilized nucleotides. Biochem J 133:499–506

Manning G, Whyte DB, Martinez R, Hunter T, Sudarsanam S (2002) The protein kinase complement of the human genome. Science 298:1912–1934

Morandell S, Stasyk T, Grosstessner-Hain K, Roitinger E, Mechtler K, Bonn GK, Huber LA (2006) Phosphoproteomics strategies for the functional analysis of signal transduction. Proteomics 6:4047–4056

Morphy R, Kay C, Rankovic Z (2004) From magic bullets to designed multiple ligands. Drug Discov Today 9:641–651

Moser K, White FM (2006) Phosphoproteomic analysis of rat liver by high capacity IMAC and LC-MS/MS. J Proteomc Res 5:98–104

Nuhse TS, Peck SC (2006) Peptide-based phosphoproteomics with immobilized metal ion chromatography. Methods Mol Biol 323:431–436

Ong SE, Blagoev B, Kratchmarova I, Kristensen DB, Steen H, Pandey A, Mann M (2002) Stable isotope labeling by amino acids in cell culture, SILAC, as a simple and accurate approach to expression proteomics. Mol Cell Proteomics 1:376–386

Ong SE, Mann M (2005) Mass spectrometry-based proteomics turns quantitative. Nat Chem Biol 1:252–262

Pinkse MW, Uitto PM, Hilhorst MJ, Ooms B, Heck AJ (2004) Selective isolation at the femtomole level of phosphopeptides from proteolytic digests using 2D-NanoLC-ESI-MS/MS and titanium oxide precolumns. Anal Chem 76:3935–3943

Rappsilber J, Ryder U, Lamond AI, Mann M (2002) Large-scale proteomic analysis of the human spliceosome. Genome Res 12:1231–1245

Ross PL, Huang YN, Marchese JN, Williamson B, Parker K, Hattan S, Khainovski N, Pillai S, Dey S, Daniels S, Purkayastha S, Juhasz P, Martin S, Bartlet-Jones M, He F, Jacobson A, Pappin DJ (2004) Multiplexed protein quantitation in Saccharomyces cerevisiae using amine-reactive isobaric tagging reagents. Mol Cell Proteomics 3:1154–1169

Rush J, Moritz A, Lee KA, Guo A, Goss VL, Spek EJ, Zhang H, Zha XM, Polakiewicz RD, Comb MJ (2005) Immunoaffinity profiling of tyrosine phosphorylation in cancer cells. Nat Biotechnol 23:94–101

Schmidt A, Kellermann J, Lottspeich F (2005) A novel strategy for quantitative proteomics using isotope-coded protein labels. Proteomics 5:4–15

Szardenings K, Li B, Ma L, Wu M (2004) Fishing for targets: novel approaches using small molecule baits. Drug Discovery Today: Technologies 1:9–15

Vella F, Ferry G, Delagrange P, Boutin JA (2005) NRH:quinone reductase 2: an enzyme of surprises and mysteries. Biochem Pharmacol 71:1–12

Wang Y, Li R, Du D, Zhang C, Yuan H, Zeng R, Chen Z (2006) Proteomic analysis reveals novel molecules involved in insulin signaling pathway. J Proteome Res 5:846–855

Weisberg E, Manley PW, Cowan-Jacob SW, Hochhaus A, Griffin JD (2007) Second generation inhibitors of BCR-ABL for the treatment of imatinib-resistant chronic myeloid leukaemia. Nat Rev Cancer 7:345–356

Wissing J, Godl K, Brehmer D, Blencke S, Weber M, Habenberger P, Stein-Gerlach M, Missio A, Cotten M, Muller S, Daub H (2004) Chemical proteomic analysis reveals alternative modes of action for pyrido[2,3-d]pyrimidine kinase inhibitors. Mol Cell Proteomics 3:1181–1193

Yang TT, Xiong Q, Graef IA, Crabtree GR, Chow CW (2005) Recruitment of the extracellular signal-regulated kinase/ribosomal S6 kinase signaling pathway to the NFATc4 transcription activation complex. Mol Cell Biol 25:907–920

Zhang Y, Wolf-Yadlin A, Ross PL, Pappin DJ, Rush J, Lauffenburger DA, White FM (2005) Time-resolved mass spectrometry of tyrosine phosphorylation sites in the epidermal growth factor receptor signaling network reveals dynamic modules. Mol Cell Proteomics 4:1240–1250

Zieske LR (2006) A perspective on the use of iTRAQ reagent technology for protein complex and profiling studies. J Exp Bot 57:1501–1508

Ernst Schering Foundation Symposium Proceedings, Vol. 3, pp. 29–41
DOI 10.1007/2789_2007_061

Published Online: 18 December 2007

PKC Isotype Functions in T Lymphocytes

G. Baier(✉)

Medical University of Innsbruck, Schoepfstraße 41, 6020 Innsbruck, Austria
email: *Gottfried.Baier@uibk.ac.at*

Abstract. The main function of mature T cells is to recognize and respond to foreign antigens by a complex activation process involving differentiation of the resting cell to a proliferating lymphoblast actively secreting immunoregulatory lymphokines or displaying targeted cytotoxicity, ultimately leading to recruitment of other cell types and initiation of an effective immune response. In order to understand the physiology and pathophysiology of T lymphocytes, it is necessary to decode the biochemical processes that integrate signals from antigen, cytokine, integrin and death receptors. The principal upon which our work is based is to explore and identify gene products of distinct members of the AGC family of protein serine/threonine kinases as key players mediating cell growth regulation. Given the established important role of PKC θ as regulator of T cell fate and knowing that several other PKC isotypes are also expressed in T cells at a high level, we now summarize the physiological and non-redundant functions of PKC α, β, δ, ε, ζ and θ isotypes in T cells. This review describes the current knowledge of the physiological and non-redundant functions of the PKC gene products in T cells.

1 Introduction

1.1 The PKC Kinases, a Gene Family of Nine Isotypes

Protein kinase C (PKC) are members of the serine/threonine protein kinase family, originally identified by Nishizuka and colleagues in 1977 as a cyclic nucleotide-independent protein kinase that phosphorylated histone and protamine in bovine cerebellum (Yamamoto et al. 1977). According to their structure and cofactor dependency PKC can be classified into conventional PKC α, β, γ, novel PKC δ, ε, θ, η and atypical PKC ζ, ι (Fig. 1). Conventional PKC require Ca^{2+} and diacylglycerol (DAG) for activation, novel PKC are Ca^{2+}-independent and atypical PKC require neither Ca^{2+} nor DAG for activation. Some of the isoenzymes exist as different splice variants (for a review see Kofler et al. 2003).

PKC became the focus of attention among cellular biologists interested in signal transduction after it was discovered that it is activated by the inositol phospholipid-derived second messenger, DAG and by phorbol esters and other tumour promoters. Members of the PKC family of serine/threonine protein kinases have been implicated in numerous cellular responses, such as cell growth and differentiation, cell cycle control, homeostasis, synaptic transmissions, the activation of ion fluxes, secretion and tumorigenesis in a wide range of cell types and tissues. *In vitro*, PKC can phosphorylate multiple protein substrates, including receptors and other membrane proteins, contractile and cytoskeletal proteins and enzymes. Like many other signalling effectors PKC is not a single entity but product of the nine mammalian PKC genes with distinct chromosomal locations (see Kofler et al. 2003). The reasons for the heterogeneity of PKC isotypes are not yet fully understood. Expression of more than a single PKC isotype in a given cell could implicate functional redundancy and/or functional specialization.

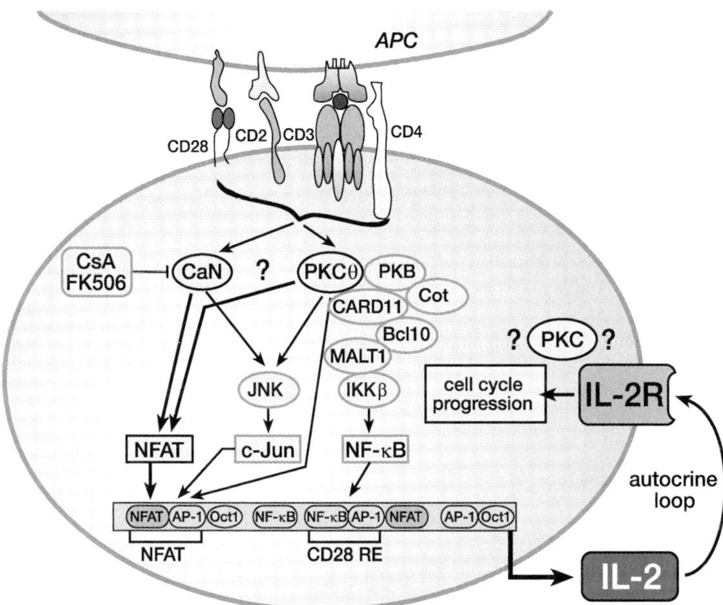

Fig. 1. A schematic representation of the physiological and non-redundant PKC gene functions in T cells: PKC θ and most likely also PKC α are critical to antigen receptor mediated T cell activation. In contrast, the isotype selective functions of PKC β, δ, ε and ζ appear to be dispensable in signalling pathways leading to proliferation and IL2 secretion of primary $CD3^+$ T cells. Description of these suspected PKC gene functions is given in the text

T lymphocytes contain up to eight different species of PKC isotypes (see Baier 2003), which makes it rather difficult to determine the specific cellular functions of these individual enzymes. Additionally, some confusion has risen from the use of phorbol esters as pleiotropic activator of PKC isotypes. This may be related to a differential regulation of PKC isotypes by phorbol esters. Phorbol esters also target receptors other than PKC, a concept that has been largely ignored in the past. These novel non-PKC phorbol ester receptors include chimaerins (a family of Rac-GTPase-activating proteins), Unc-13/Munc-13 (a family of scaffold proteins involved in exocytosis), RasGRP (a Ras exchange factor family), Myotonic Dystrophy Kinase-Related Cdc42-Binding Kinases (MRCK) α/β and DAG kinases β/γ. Two other PKC-related protein kinases, PKCμ and PKCν, are also activated by DAG/phorbol esters, but they contain additional functional domains and display a different pattern of substrate specificity. Therefore they have been classified separately as members of the PKD family (Rykx et al. 2003).

Nevertheless, it is well established that members of the PKC family have important functions in immune cells. Despite significant progress in assembling the PKC puzzle in immune cells, there is one drawback: most of the published studies were done in cell lines and the question arises whether these results are transferable to more physiological systems, such as primary $CD3^+$ T cells. For example, the widely used leukaemic Jurkat cell line is deficient in the lipid phosphatases Scr homology 2 domain containing inositol polyphosphate phosphatase (SHIP) and phosphatase and tensin homolog (PTEN), negative regulators of phosphatidylinositol 3-kinase (PI3 K)-signalling. Thus, the cells contain high levels of phosphatidylinositol-3,4,5-trisphosphat (PIP3) and therefore demonstrate abnormal activation of pleckstrin homology (PH) domain-containing proteins such as PDK-1, PLCγ1 or Vav-1, all established upstream regulators of the PKC family. To avoid conclusions that are rather due to the peculiarities of transformed leukaemic cell lines, we focus here mainly on PKC functions in primary $CD3^+$ T cells.

2 Regulation of PKC Activity in T Cells

The main function of mature T cells is to recognize and respond to foreign antigens by a complex activation process involving differentiation of the resting cell to a proliferating lymphoblast actively secreting immunoregulatory lymphokines or displaying targeted cytotoxicity, ultimately leading to recruitment of other cell types and initiation of an effective immune response.

Inactive PKC (when the substrate binding site in the catalytic domain is occupied intramolecularly by the pseudosubstrate motif) are phosphorylated by the phosphoinositide-dependent kinase PDK-1 in the activation loop and subsequently autophosphorylated at two sites at the C terminus (turn motif and hydrophobic motif). Nevertheless, regulation of the phosphostatus of PKC is still not yet fully understood and may differ for the different PKC subfamily members: in the Jurkat tumor T cell line, the three classical phosphorylation sites of PKC θ (in activation loop, turn motif and hydrophobic motif) are constitutively phosphorylated and only slightly induced upon receptor ligation. At least one additional, activation-induced autophosphorylation site appears to exist. In resting T cells this Thr-219 autophosphorylation site of PKC θ is phosphorylated to low stoichiometry, even frequently undetectable, suggesting that PKC θ is mostly inactive or the autophosphorylation site is the target for efficient dephosphorylation (Thuille et al. 2005). Yet another important missing link in our understanding of the role of PKC isotype (auto)phosphorylation is the lack of information on the protein phosphatase (PPase)(s) that dephosphorylate(s) these sites. Their identification will be important for determining the sequence of events that accompany catalytic activation versus suppression.

PKC function is regulated by subcellular localization, which depends on both kinase:lipid and kinase:protein interactions. PKC activation is generally associated with translocation of PKC from one cellular compartment to another, which contains lipid activators and proteins that bind the activated form of the enzyme in proximity to substrates. The binding of calcium results in translocation of the enzyme to membranes by increasing its affinity for negatively charged phospholipids (PS), and leads to the release of the pseudosubstrate domain from the catalytic site (via a conformational change). Each C2 domain thereby appears to interact with two calcium ions and multiple PS molecules. Binding of DAG than stabilizes the PKC:membrane interaction and prevents the pseudosubstrate domain from the inhibitory binding to the catalytic cleft. For the conventional and novel PKC, PDK-1 phosphorylation is constitutive, whereas for the atypical PKC, phosphorylation by PDK-1 is under partial regulation of 3′phosphoinositides providing an additional mode of atypical PKC regulation (Newton et al. 2001).

Once at the site of a particular substrate, PKC induce phosphorylation at serine/threonine residues, which in turn alters the function of the substrate protein. The established concept of PKC stimulation includes activation of phosopholipase Cγ (PLCγ), which cleaves phosphatidylinositol 4,5-bisphosphate (PIP2), a minor component of the plasma membrane, localized at the inner leaflet of the phospholipid bilayer, to generate two products: inositol trisphosphate (IP3) and DAG. Whereas DAG remains associated with the plasma membrane, the other second messenger IP3, is released into the cytosol. Interaction of IP3 with its receptors in the endoplasmic reticulum causes the release of Ca^{2+} from this intracellular storage site into the cytosol, immediately raising intracellular free Ca^{2+} levels several-fold. Depletion of the endoplasmic reticulum calcium stores triggers the opening of Ca^{2+} release-activated channels (CRAC) in the plasma membrane that now let more Ca^{2+} into the cell, thus sustaining the signal. The increased intracellular free Ca^{2+} together with DAG contributes to the activation of PKC. The effects of IP3 are mimicked by using a Ca^{2+} ionophore, such as ionomycin, which allows Ca^{2+} to move into the cytosol from the extracellular fluid. The effects of DAG can be mimicked by phorbol esters, plant products that bind, for instance, to PKC isotypes and activate them directly. Importantly, phorbol esters can replace DAG as activators, nevertheless, they also down-regulate PKC during long-term treatment.

3 Role of PKC in T Lymphocyte Physiology

The original discovery of PKC genes as major cellular receptors for phorbol ester, a pleiotropic tumour promoter, lead to the obvious conclusion that PKC participate solely in the regulation of cell growth. However, PKC are also found at high levels in post-mitotic cells such as platelets, granulocytes and neurons, suggesting that these kinases play important roles in highly specialized signalling functions including differentiation. The discoveries that hydrolysis of inositol phospholipids is triggered by engagement of the T cell receptor (TCR) and that phorbol esters together with Ca^{2+} ionophores mimic antigenic T cell stimulation suggested an important role for PKC in T cell signalling. This idea was encouraged by several studies demonstrating that PKC inhibition diminishes TCR-induced T cell response (for a review see Altman et al. 1990).

3.1 PKC α

In a recent study, by using kinase mutants, RNA interference (RNAi), and chemical inhibitors, Trushin and coworkers demonstrated that PKC α but not PKC β is required for NFκB activation following TCR/CD28-induced T cell activation in Jurkat cells. These results suggest that PKC α acts upstream of PKC θ to activate NFκB during T cell activation, although this is not confirmed for primary T cells so far (Trushin et al. 2003). PKC α is also implicated in TCR-dependent interleukin-2 (IL2) receptor expression in stimulated human lymphocytes. An anti-apoptotic role for PKC α is confirmed by Jurkat cell studies: during induction of apoptosis by Fas ligation the activity of PKC α is inhibited through activation of okadaic acid-sensitive phosphatase (PP2A) (Chen and Faller 1999).

PKC $\alpha^{-/-}$ mice are viable and developed comparably to their wild-type littermates. Recent results from our laboratory demonstrate normal development of $CD3^+$ T cells as well as normal relative distribution of $CD4^+$ and $CD8^+$ subsets in PKC α-deficient mice. Mature $CD3^+$ PKC α-deficient T cells, however, have a severe decrease in TCR induced interferon (IFN) γ production, when compared to wild-type littermates. Following immunization with a T cell-dependent antigen, the PKC α-deficient mice displayed an impaired and selective TH1-dependent IgG2a/b switch, while IgM and IgG1 responses were normal (Pfeifhofer et al. 2006). These data indicate that PKC α is part of a signalling pathway that is necessary for full antigen receptor mediated T cell activation and TH1 differentiation.

3.2 PKC β

Volkov et al. showed a role for PKC β in the migration of T cells, using a human T lymphoma line (HUT-78) and activated human peripheral blood lymphocytes (PBL). It is well known that during inflammation, T cells enter the peripheral inflamed tissues and roll along microvascular endothelial cells to sample inflammatory signals. T cell recruitment is mediated by the integrin receptor known as lymphocyte function-associated antigen-1 (LFA-1), which binds with high affinity to the counter-receptor intercellular adhesion molecule 1 (ICAM-1) expressed by endothelium at inflammatory sites. T cells arrested in the target tissues undergo cell spreading and polarization and subsequent migration across the vascular wall to seek out their extravascular targets in the inflamed tissue. LFA-1-induced signalling is demonstrated to recruit PKC βI and PKC δ to microtubules (Volkov et al. 1998). Furthermore, expression studies showed that PKC βI promoted cell polarization and enhanced motility of an otherwise defective PKC β-deficient T cell clone (Volkov et al. 2001).

Two independent scientific groups indicated a specific function of PKC β for regulation of TCR-CD28 induced-signalling, IL2 gene transcription and IL2 secretion from T cell lines: The inhibition of the PKC β synthesis in stimulated Jurkat cells treated with specific phosphorothiothiate-modified antisense oligonucleotides resulted in suppression of phosphorylation and activation of Raf1, MEK and ERK and led to a marked reduction of IκBα phosphorylation, a significant inhibition of NFAT activation and a complete suppression of IL2 transcription and secretion. A similar study with a PKC β-deficient clone of HUT 78 cells demonstrated that PKC β is not necessary for transcription and translation of the IL2 gene but only for the cellular export of IL2 molecules from T cells via exocytosis (Long et al. 2001). The difference in the IL2 synthesis between the two studies could be explained by a cell-type specific PKC gene expression and function. A role for PKC β in the regulation of Ca^{2+} entry in Jurkat T cells is shown by Haverstick et al. when electroporation of an anti-PKC βI antibody led to an increase in the rate of Ca^{2+} influx following T cell stimulation (Haverstick et al. 1997)

PKC $\beta^{-/-}$ mice, however, have an inconspicuous T cell phenotype. PKC β-deficient T cells showed normal proliferative responses to CD3/CD28 stimulation as well as to activation via allogenic MHC. Consistently, IL2 production and secretion is normal and activation-induced expression of the surface markers CD44, CD69 and CD25 is comparable to wild-type levels. T cell development, as reflected by numbers of $CD3^{+}$ cells and $CD4^{+}$/$CD8^{+}$ distribution, seemed to be unaffected by loss of PKC β function (Thuille et al. 2004). These data suggest that PKC β is dispensable in T cell signalling (in strict contrast to B-cells, Leitges et al., 1996). However, it is conceivable that other members of the PKC family can compensate for the lack of PKC β in T cells.

3.3 PKC δ

Interestingly, PKC δ-deficient T cells proliferated more strongly than wild-type controls when stimulated with allogenic MHC *ex vivo*. Consistently, CD3 application leads to an enhanced IL2 secretion *in vivo*. These results suggest that PKC δ is a negative regulator of T cell activation. Interestingly, the hyper-responsiveness of PKC $\delta^{-/-}$ T cells is not observed in the proliferation assay, in which TCR/CD3 are cross-linked by plate-bound CD3 antibodies. This points to an additional factor that is required for a pronounced PKC δ effect and that is only provided in more physiological systems such as cell-bound allogenic MHC (Gruber et al. 2005).

3.4 PKC ζ

To address the role of PKC ζ in T cell activation, San Antonio and coworkers have generated Jurkat T cell transfectants expressing either the wild-type or dominant-negative mutant of PKC ζ; they thus observed that PKC ζ regulates the transactivation of NFAT in Jurkat T cells. The authors propose that phosphorylation of the N-terminal transactivation domain of NFAT by PKC ζ might mediate modulation of gene expression (San Antonio et al. 2002).

We identified PKC ζ as a PKC θ interacting protein by a yeast two-hybrid screen; using an over-expresion approach we could demonstrate that PKC ζ is functionally involved in PKC θ-mediated activation of NFκB in Jurkat T cells. PKC ζ is recruited to the plasma membrane lipid raft microdomains of T cells, where a physical and functional interaction with PKC θ may be biologically relevant. To further evaluate its non-redundant physiological role in T cell activation we used our established PKC ζ knock out mouse line: however, PKC ζ-deficient peripheral T cells responded normally to CD3/CD28 activation stimuli, indicating a functional redundancy with other PKC family members, most likely with PKC ι (Gruber et al. 2007).

3.5 PKC θ

PKC θ was found to be selectively expressed in lymphoid organs, skeletal muscle and the nervous system (for a review see Baier 2003). PKC $\theta^{-/-}$ mice are viable and developed normally when compared to their littermates. T cell development of $CD3^+$ as well as the relative distribution of $CD4^+$ and $CD8^+$ subsets is normal in PKC θ-deficient mice. PKC θ-deficient T cells showed decreased proliferation and IL2 secretion upon stimulation with CD3 and costimulation with CD3/CD28 compared to wild-type cells. Their response to allogenic MHC in the mixed lymphocyte reaction is also significantly impaired. EMSA analysis of the transcription factors important for IL2 expression revealed a decreased activation of NFκB and AP-1 as well as an almost complete abrogation of NFAT activation upon costimulation with CD3/CD28 (Pfeifhofer et al. 2003). This is in contrast to a previous study (Sun et al. 2000) that stated a normal NFAT activation in PKC γ-deficient T lymphocytes (these cells are generated by inserting a neomycin cassette, leaving a residual open reading frame of 365 amino acids). Consistent with our findings, IP3 generation and Ca^{2+} mobilization are also diminished in PKC $\theta^{-/-}$ T cells. These results suggest a role of PKC θ upstream of PLC γ. Since PLC γ activation in turn generates DAG, a positive regulator of PKC θ, these two enzymes could be part of a positive feedback loop. The expression of activation-induced surface markers CD44, CD69 and CD25 (the α chain of the IL2 receptor) appeared unaffected by the complete loss of function of PKC θ.

In antigen receptor-induced NFκB activation, Carma1 (also known as caspase recruitment domain [CARD]11), a CARD-containing membrane-associated guanylate kinase family protein, was shown to play an essential role. Carma1 is essential for antigen-induced IL2 and IFNγ production, but dispensable for proliferation of T cells. Carma1 controls entry of IKK into lipid raft aggregates and the central region of the immune synapse, as well as activation of IKK downstream of PKC θ (Hara et al. 2003). In this regard, TCR/CD3 ligation is sufficient to induce recruitment and activation of PKC θ. There is a selective complex formation and cooperation of PKC θ and Akt1/PKB α in the NFκB transactivation cascade and Akt1/PKB α provides the CD28 costimulatory signal for up-regulation of IL2 and IFNγ via activating a NFκB in T cells (Kane et al. 2001).

Taken together, PKC θ is the first identified member of the PKC family to play a critical role in the Ca^{2+}/NFAT, AP-1 and NFκB pathway to activate the IL2 promoter. The reduced Ca^{2+} mobilization in PKC $\theta^{-/-}$ T cells might further enhance the observed partial reduction of NFκB and AP-1 activation, since both signalling pathways are known to be dependent on the amplitude and duration of the Ca^{2+} signal.

As a recent finding, PKC θ appears to be required for the development of a robust lung inflammatory response controlled by TH2 cells in vivo: PKC $\theta^{-/-}$ mice are protected from pulmonary allergic hypersensitivity responses such as airway hyperresponsiveness, eosinophilia, and immunoglobulin E production to inhaled allergen. Furthermore, TH2 cell immune responses against *Nippostrongylus brasiliensis* are severely impaired in PKC $\theta^{-/-}$ mice. In striking contrast, PKC $\theta^{-/-}$ mice mounted protective TH1 immune responses and are resistant against infection with *Leishmania major*. Consistently, PKC θ plays a lesser role in the development of a similar lung inflammatory response controlled by TH1 cells: PKC θ-deficient mice are able to mount a TH1-mediated lung inflammatory response (Salek-Ardakani et al. 2004; Marsland et al. 2004). In summary, these results show that PKC θ is critical for the development of *in vivo* TH2 cell immune responses. These data highlight PKC θ as an important target in therapy against TH2 cell-related diseases.

4 Conclusion

In T cells, both PKC θ and PKC α appear to take part in signalling pathways that are necessary for full antigen receptor mediated T cell activation and T lymphocyte immunity. PKC θ is critical for the Ca^{2+}/NFAT, AP-1 and NFκB transacticvation downstream of the antigen receptor and PKC α appears essential for IFNγ expression and TH1 dependent immune responses. In contrast, PKC β, δ, ε and ζ appear to be dispensable in antigen receptor-induced activation of T cells.

References

Altman A, Coggeshall KM, Mustelin T (1990) Molecular events mediating T cell activation. Adv Immunol 48:227–360

Baier G (2003) The PKC gene module: molecular biosystematics to resolve its T cell functions. Immunol Rev 192:64–79

Chen CY, Faller DV (1999) Selective inhibition of protein kinase C isozymes by Fas ligation. J Biol Chem 274:15320–15328

Gruber T, Barsig J, Pfeifhofer C et al. (2005) PKCdelta is involved in signal attenuation in CD3(+) T cells. Immunol Lett 96:291–293

Gruber T, Thuille N, Fresser F et al. (2007) PKCtheta T cell signaling function cooperates with aPKCzeta. Mol Immunol (in press)

Hara H, Wada T, Bakal C et al. (2003) The MAGUK family protein CARD11 is essential for lymphocyte activation. Immunity 18:763–775

Haverstick DM, Dicus M, Resnick MS, Sando JJ, Gray LS (1997) A role for protein kinase CbetaI in the regulation of Ca2+ entry in Jurkat T cells. J Biol Chem 272:15426–15433

Kane LP, Andres PG, Howland KC, Abbas AK, Weiss A (2001) Akt provides the CD28 costimulatory signal for up-regulation of IL-2 and IFN-gamma but not TH2 cytokines. Nat Immunol 2:37–44

Kofler K, Erdel M, Utermann G, Baier G (2002) Molecular genetics and structural genomics of the human protein kinase C gene module. Genome Biol 3:R14–16

Leitges M, Schmedt C, Guinamard R et al. (1996) Immunodeficiency in protein kinase cbeta-deficient mice. Science 273:788–791

Long A, Kelleher D, Lynch S, Volkov Y (2001) Cutting edge: protein kinase C beta expression is critical for export of Il-2 from T cells. J Immunol 167:636–640

Marsland BJ, Soos TJ, Spath G, Littman DR, Kopf M (2004) Protein kinase C theta is critical for the development of in vivo T helper (Th)2 cell but not Th1 cell responses. J Exp Med 200:181–189

Newton AC (2001) Protein kinase C: structural and spatial regulation by phosphorylation, cofactors, and macromolecular interactions. Chem Rev 101:2353–2364

Pfeifhofer C, Gruber T, Letschka T, Thuille N, Lutz-Nicoladoni C, Hermann-Kleiter N, Braun U, Leitges M, Baier G (2006) Defective IgG2a/2b class switching in PKC alpha$^{-/-}$ mice. J Immunol 176:6004–6011

Pfeifhofer C, Kofler K, Gruber T et al. (2003) Protein kinase C theta affects Ca2+ mobilization and NFAT cell activation in primary mouse T cells. J Exp Med 197:1525–1535

Rykx A, De Kimpe L, Mikhalap S et al. (2003) Protein kinase D: a family affair. FEBS Lett 546:81–86

Salek-Ardakani S, So T, Halteman BS, Altman A, Croft M (2004) Differential regulation of Th2 and Th1 lung inflammatory responses by protein kinase C theta. J Immunol 173:6440–6447

San Antonio B, Iniguez MA, Fresno M (2002) Protein kinase Czeta phosphorylates nuclear factor of activated T cells and regulates its transactivating activity. J Biol Chem 277:27073–27080

Sun Z, Arendt CW, Ellmeier W et al. (2000) PKC-theta is required for TCR-induced NF-kappaB activation in mature but not immature T lymphocytes. Nature 404:402–407

Thuille N, Gruber T, Bock G, Leitges M, Baier G (2004) Protein kinase C beta is dispensable for TCR-signaling. Mol Immunol 41:385–390

Thuille N, Heit I, Fresser F, Krumböck N, Bauer B, Leuthaeusser S, Dammeier S, Graham C, Copeland TD, Shaw S, Baier G (2005) Critical role of novel Thr-219 autophosphorylation for the cellular function of PKCtheta in T lymphocytes. EMBO J 16:3869–3880

Trushin SA, Pennington KN, Carmona EM et al. (2003) Protein kinase Calpha (PKCalpha) acts upstream of PKCtheta to activate IkappaB kinase and NF-kappaB in T lymphocytes. Mol Cell Biol 23:7068–7081

Volkov Y, Long A, Kelleher D (1998) Inside the crawling T cell: leukocyte function-associated antigen-1 cross-linking is associated with microtubule-directed translocation of protein kinase C isoenzymes beta(I) and delta. J Immunol 161:6487–6495

Volkov Y, Long A, McGrath S, Ni ED, Kelleher D (2001) Crucial importance of PKC-beta(I) in LFA-1-mediated locomotion of activated T cells. Nat Immunol 2:508–514

Yamamoto M, Takai Y, Hashimoto E, Nishizuka Y (1997) Intrinsic activity of guanosine 3′,5′-monophosphate-dependent protein kinase similar to adenosine 3′,5′-monophosphate-dependent protein kinase. I. Phosphorylation of histone fractions. J Biochem (Tokyo) 81:1857–1862

Ernst Schering Foundation Symposium Proceedings, Vol. 3, pp. 43–61
DOI 10.1007/2789_2007_070

Published Online: 29 February 2008

Systems Biology of T Cell Activation

J.A. Lindquist(✉), B. Schraven

Institute of Molecular and Clinical Immunology, Otto-von-Guericke University, Leipziger Strasse 44, 39120 Magdeburg, Germany
email: *Jon.Lindquist@med.ovgu.de*

Abstract. T lymphocytes are central players in the adaptive immune response to pathogens. Cytotoxic T cells are able to identify and eliminate virally infected cells, while helper T cells support B lymphocyte-dependent antibody production as well as produce the cytokines that will determine whether a cell or antibody-mediated immune response is required. The activation of T cells by pathogens is a complex process requiring multiple tightly regulated signaling pathways. Defects within this network, however, can cause severe and chronic disorders such as autoimmunity. Therefore, improving our understanding of how T cells discriminate between antigens and how these signals are organized to yield distinct immune responses is of importance as this may lead to the identification of novel drug targets and better therapeutic strategies.

1 Introduction

The immune system has evolved to protect us from invading pathogens (e.g., bacteria or viruses) and thus must be able to recognize a wide variety of foreign antigens (i.e., those substances that generate an immune response). However, the system is not centralized in one organ, but is distributed throughout the body, beginning with the physical barriers, such as skin and mucosa, that separate us from our environment. It contains both primary immune organs, such as the bone marrow and thymus, which are responsible for generating and educating the cells that make up the immune system, and secondary immune organs, such as lymph nodes, where specific immune responses are initiated. The immune system is composed of two components, innate and adaptive immunity, both of which possess the ability to recognize and eliminate pathogens. The difference is that an innate response is always the same, whereas the adaptive immune response is specific and improves with each antigenic encounter. It is these features, (1) specificity, (2) the ability to distinguish self from other, and (3) memory, that set the two responses apart from one another. Both have cellular and humoral components. The cells of the immune system are all generated from a single pluripotent stem cell that gives rise to both lymphoid and myeloid precursors. While the myeloid precursor differentiates into the cells of the innate immune system, the lymphoid precursor forms the cells that make up the adaptive immune system. These are primarily B and T lymphocytes. While both recognize antigens using membrane-bound "antigen receptors," they do so in completely different ways. B cells utilize membrane-bound immunoglobulin [i.e., the B cell receptor (BCR)] and are capable of recognizing a wide variety of antigens (including proteins, lipids, carbohydrates, and nucleic acids). T cells, on the other hand, use the T cell receptor (TCR) that recognizes peptide antigens only when they are presented in the context of an MHC molecule.

A typical immune response occurs something like this: A pathogen enters the body and begins to multiply. As the number of invaders increases, the innate immune system is activated and begins to fight the infection. This in turn leads to the recruitment of phagocytes, such as macrophages or dendritic cells (or both), which take up the pathogens and digest them. In this way, some protein antigens from the pathogen are processed, loaded onto MHC molecules, and presented on the surface of the phagocytes to T cells. Under the right conditions, when the T cell encounters a peptide–MHC complex for which it has a high enough affinity, this can lead to its activation, the end result of which is that this single antigen-specific T cell proliferates (increasing in numbers exponentially) and differentiates into both effector and memory populations. The effector cells then aid in eliminating the pathogen and will eventually themselves be killed via mechanisms that maintain homeostasis within the immune system. The memory cells, meanwhile, are long-lived and, should they encounter the same antigen again, will be able to respond much more quickly. If daughter cells are genetically identical, then how can two cell populations arise from a single antigen-specific T cell? A recent study answered this question by showing that activation [i.e., contact with an antigen-presenting cell (APC)] polarizes the T cell, resulting in an asymmetric division of the activated T lymphocyte, which in turn gives rise to the two populations (Chang et al. 2007). T cells that are in contact with the APC (i.e., proximal to the site of contact) are enriched in TCR and many signaling molecules, thus giving rise to the effector population, while the distal daughter cells give rise to the memory population.

2 Antigen Presentation

The MHC is one of the most gene-rich clusters within the genome. It possesses more than 100 genes relevant to the function of the immune system, including those genes that encode the MHC molecules. This includes the genes for both the classical MHC class I alleles (i.e., HLA-A, -B, and -C) as well as those for the MHC class II molecules (HLA-DR, -DQ, and -DP). Class I molecules are composed of a single membrane bound heavy chain, which folds around a soluble light chain known as β_2-microglobulin (β_2m). MHC class II molecules, on the other hand, are composed of two equally sized membrane-bound chains called α and β. Class I molecules are synthesized and assembled within the endoplasmic reticulum (ER), where they bind peptides of 8 to 11 amino acids in length, which are primarily, but not exclusively, derived from the degradation of intracellular proteins by a proteolytic protein complex called the proteasome. Since these antigens are derived from intracellular proteins, this system is optimized for the detection of those pathogens that seek to hide within our cells, such as viruses or intracellular bacteria. MHC class I molecules are known to possess binding pockets or anchor positions, which typically prefer to bind a single amino acid. However, in a peptide of nine amino acids, where two such anchor positions exist, this still leaves seven positions "free" to be filled by any of the 20 amino acids. In this scenario, such an MHC molecule could bind as many as 1,280,000,000 different peptides (i.e., 20^7). Given that a cell might express as many as six different class I molecules, we then have the ability to present close to 10 billion different peptides.

This number is somewhat larger for class II molecules for two reasons. First, the possible number of class II molecules on a cell is greater, since they are composed of both an α and a β chain for which multiple alleles exist, so some mixing and matching occurs between the proteins. And also, class II molecules present larger peptides than class I molecules, typically 12 to 25 amino acids in length that are derived from extracellular antigens. Thus, MHC class II molecules are almost exclusively found on phagocytic cells and present antigens derived from either extracellular bacteria or parasites. However, there also exists the possibility that phagocytosed material can enter into the cytosol and be loaded onto MHC class I molecules. This process is called "cross-presentation" and is thought to be important in initiating immune responses to intracellular pathogens. Remember that an adaptive immune response is generally initiated within the secondary lymphoid organs, such as the lymph nodes, and thus if a person has a viral infection in the periphery (i.e., within the mucosa), antigens must somehow be transported from the site of infection to the lymph nodes so that the alarm can be raised and an appropriate defensive response mounted to fight off the invaders.

3 Central Tolerance

Naturally, the cells presenting antigens are not biased toward which peptides they present and thus they present both self-antigens as well as foreign. So, how is it that T cells are not constantly being activated against

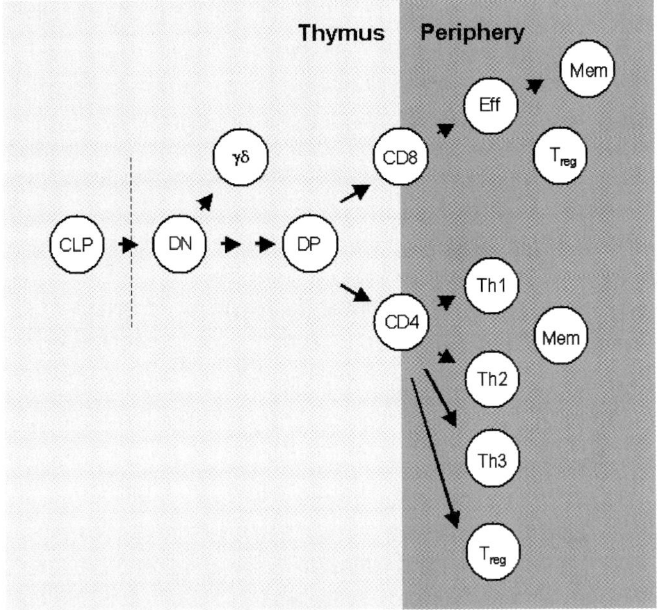

Fig. 1. T cell development. The developmental pathway for T cells is depicted here, beginning with the common lymphoid precursor (CLP). Development within the thymus has been simplified to indicate that γδ T cells develop independently of the αβ lineage. Double negative (DN) cells express neither CD4 nor CD8 while double positive (DP) cells express both and will give rise to both the naive $CD4^+$ and $CD8^+$ populations that exit the thymus and enter the periphery. Other abbreviations: *Eff*, effector; *Mem* memory; *Th*, T helper; *Treg*, regulatory T cell

self-antigens? One reason is certainly due to central tolerance, which occurs in the thymus, the primary immune organ responsible for educating T cells. Here, thymic epithelial cells display self-antigens. And as T cell precursors develop (see Fig. 1), they are screened against a panel of self-antigens. Those T cells that fail to properly assemble their antigen receptor (i.e., the TCR) die of neglect, since they fail to receive a necessary survival signal, while those cells that bind with too high an affinity to self-antigens are negatively selected and thus removed from

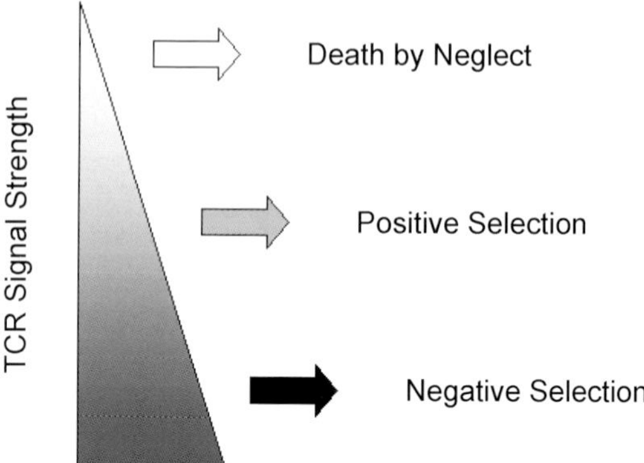

Fig. 2. Thymic selection. T cell progenitors develop within the thymus. Those that fail to correctly assemble a TCR die by neglect, while those cells that successfully express a TCR survive. These cells then undergo a second round of selection upon cells expressing self-peptide:MHC complexes. Those which bind with too high an affinity for themselves are deleted (i.e., negatively selected), so that only those cells with a weak or no affinity for self survive

the T cell repertoire (see Fig. 2). Therefore, only those cells that successfully assemble a TCR with a weak affinity for themself survive and can exit into the periphery, which is in the order of a few million T cells per day. In total there are approximately 250 million T cells distributed throughout the human body.

4 TCR Structure

The T cell receptor is composed of two chains that are capable of binding to antigen (see Fig. 3). The majority of T cells possess αβ TCRs, although a minor subset of T cells exist that bear the combination γδ. It is important to remember that although a T cell expresses nearly 100,000 TCRs upon its surface, each of these receptors is identical and possesses the same specificity for antigen. In addition to the chains that interact with antigen, the TCR is also composed of four additional molecules called the CD3 complex; since this complex is unique to T cells, it is often used as a marker to identify T cells. The CD3 complex consists

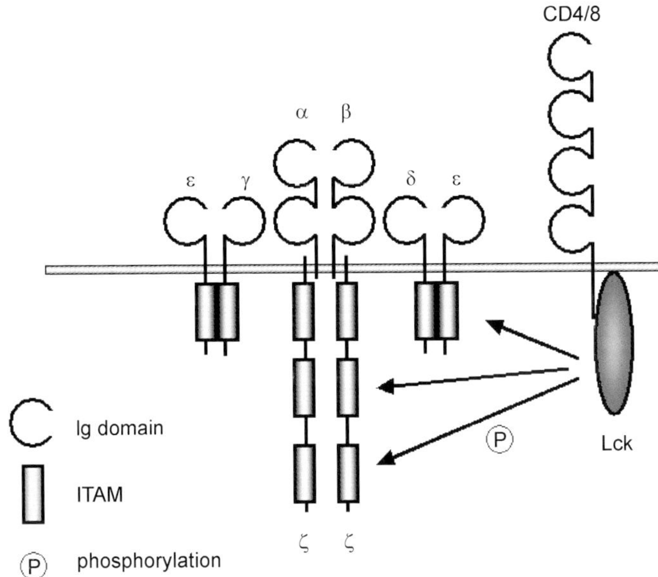

Fig. 3. Structure of the T cell receptor. The receptor used by T cells to recognize antigen (i.e., peptide:MHC complexes) is the T cell receptor, which is composed of two chains (α and β) that together form the antigen-binding site and several associated molecules (γ, δ, ϵ, and ζ) that are responsible for transducing the signal. The same antigen complex bound by the TCR is also recognized by the coreceptor CD4 or CD8. Associated with the cytoplasmic tail of the coreceptor is a member of the Src family of protein tyrosine kinases, $p56^{Lck}$, which, when recruited into the complex with TCR and peptide:MHC, phosphorylates specific motifs located within the cytoplasmic tails of the receptor-associated molecules that are necessary for activation called immunoreceptor tyrosine-based activation motifs (ITAMs). Since several of these molecules belong the immunoglobulin (Ig) superfamily, the Ig domains are indicated

of two heterodimers: γϵ and δϵ. Additionally, the TCR possesses a homodimer of ζ chains (CD247). The stoichiometry of the TCR complex is proposed to be $\alpha\beta(\gamma\delta\epsilon_2\zeta_2)$ although dimeric TCRs $[(\alpha\beta)_2(\gamma\delta\epsilon_2\zeta_2)]$ have been demonstrated (San Jose et al. 1998; Fernández-Miguel et al. 1999). Additionally, clustering of TCRs has recently been proposed to play an important role in the activation process (Bunnell et al. 2006; Varma et al. 2006; Minguet et al. 2007).

5 T Cell Activation

T cells within the body are not a homogeneous population but can be divided into two populations depending upon the type of coreceptor they express: helper T cells (Th) bear the coreceptor CD4, while cytotoxic T cells (CTLs) express CD8. These populations can then be further subdivided into effector, memory, or regulatory populations based on the expression of additional surface markers. As the name implies, the coreceptors (CD4 and CD8) bind the same ligand as the T cell receptor (TCR). However, rather that being peptide specific, the coreceptors recognize conserved regions within the MHC molecules themselves. CD4 binds to MHC class II, while CD8 recognizes MHC class I. Each coreceptor has associated with its cytoplasmic tail a protein tyrosine kinase belonging to the Src family, called $p56^{Lck}$ (see Fig. 3). Thus, T cell activation is envisioned to involve a stable binding of a peptide–MHC complex to the TCR, which then allows the coreceptor to also bind to the MHC molecule, thus bringing the kinase into proximity with the TCR. Kinases are enzymes that catalyze the transfer of the terminal phosphate group from ATP to their substrate. In this case, the substrates of Lck are specific tyrosine residues located within the cytoplasmic tails of the TCR complex known as immunoreceptor tyrosine-based activation motifs (ITAMs). ITAMs are highly conserved sequences that possess two tyrosine residues. The TCR itself possesses, in total, 10 such motifs. Thus, phosphorylation of these motifs by Lck can potentially create a large number of binding sites for proteins possessing modular binding domains. One such structure is the Src homology 2 (SH2) domain. Since ITAMs are dual tyrosine motifs, it is of no surprise that the major protein that is recruited also possesses tandem SH2 domains; note that each SH2 domain binds one phosphotyrosine (pY) residue. This protein is itself a protein tyrosine kinase called the zeta-associated protein of 70 kDa (ZAP-70).

Upon binding to the phosphorylated ITAMs, ZAP-70 is itself phosphorylated by Lck, leading to its activation. Although the TCR possesses ten ITAMs, there are limited reports of other molecules being recruited to these structures, and it is also not clear how many of the ITAMs actually become phosphorylated upon activation, although there appears to be a hierarchy to their phosphorylation (Osman et al. 1996). A number of other enzymes, e.g., phosphoinositol-3-kinase (PI3K), become activated upon TCR ligation, but whether ITAMs play a direct role in their activation process is still unclear.

Normally, the binding of peptide–MHC to the TCR is not sufficient to induce full activation of a T cell, and therefore additional signals are required via costimulatory molecules. The best characterized of these molecules is CD28. Its ligands are CD80 and CD86, both of which are expressed only on professional antigen presenting cells (i.e., phagocytic cells), thus emphasizing their importance in initiating an adaptive immune response. The primary signal given by CD28 is the activation of PI3K, which is required to activate many other proteins, including PDK1, PKB/Akt, and other enzymes possessing domains that bind phospholipids. Recall that PI3K is responsible for the conversion of phosphoinositol-4,5-bisphosphate (PIP_2) into phosphoinositol-3,4,5-trisphosphate (PIP_3). Additionally, CD28 contributes to the activation of Vav1; however, the mechanism behind this is unclear at present. An additionally important critique is that much of our knowledge of CD28 signaling has been gained from studies using monoclonal antibodies. Therefore, how much of our knowledge can be applied to stimulation with endogenous ligands remains to be demonstrated.

TCR ligation activates both Src kinases and ZAP-70. Activated ZAP-70 in turn phosphorylates the linker for activation of T cells (LAT). LAT is the central molecule in T cell activation and the point at which the input signal (peptide–MHC binding) is split to activate multiple signaling cascades. The reason for this is that LAT is a transmembrane adaptor or scaffold and serves as a platform for organizing and initiating signaling complexes. Although LAT possesses multiple tyrosine residues, only the four distal tyrosines appear critical for its function. The first of these is the binding site for phospholipase Cγ1 (PLCγ1). The next two are binding sites for the complex GADS–SLP76. The latter is also a substrate of ZAP-70 and when phosphorylated recruits a member of the Tec family kinases, either Itk or Rlk, which in turn activates PLCγ1. Additionally, LAT recruits the Grb2–SOS complex

(Son of Sevenless), which is a Ras–GEF (guanine nucleotide exchange factor). The multiprotein complex consisting of PLCγ1–pLAT–GADS–SLP76–Itk is often referred to as the calcium initiation complex, since activation of PLCγ1 results in the hydrolysis of PIP_2 into inositol trisphosphate (IP_3) and diacylglycerol (DAG). IP_3 binds to the IP_3 receptor on the ER, which results in the release of intracellular calcium stores, which in turn can stimulate the membrane calcium release-activated calcium (CRAC) channels, allowing extracellular calcium into the cells. This in turn leads to activation of the phosphatase, calcineurin, which dephosphorylates the nuclear factor of activated T cells (NFAT), allowing its translocation into the nucleus.

DAG, on the other the other hand, binds to and activates protein kinase C θ (PKCθ) and RasGRP, a GEF for the Ras family of small membrane G-proteins. Activation of Ras is thought to be responsible for the activation of the mitogen-activated protein (MAP) kinase cascade, which results in the activation of the extracellular signal-regulated kinases (ERK) 1 and 2. Additionally, the LAT–GADS–SLP76 complex also recruits factors such as Hpk1 and Vav1, which contribute to the activation of p38 and Jun N-terminal kinase (JNK). However, an alternative pathway of p38 activation has been demonstrated in T cells, which involves Lck, ZAP-70, and a member of the MAGUK family, Dlgh1, which also appears to have scaffolding functions. Activation of the MAP kinases results in the formation of the Fos/Jun complex AP-1.

Furthermore, PKCθ initiates a signaling cascade via the complex CARMA-1(Card11)/Bcl10/MALT1 that results in the activation of the nuclear factor-κB (NF-κB) by inducing the degradation of the inhibitor of κB (IκB). Together, these three factors (NFAT, AP-1, and NF-κB) contribute to the transcription of many genes that are associated with T cell activation [e.g., interleukin-2 (IL-2)].

In addition to activating Src kinases, ligation of the TCR has recently been shown to have an additional effect, which appears to actually precede Src kinase activation. TCR ligation results in the induction of a conformation change within one subunit of the TCR, CD3ϵ (Gil et al. 2002). The conformational change has been shown to expose a proline-rich region within CD3ϵ that recruits the cytosolic adaptor Nck via the first of its three Src homology 3 (SH3) domains. Although it is unclear at this time what purpose recruiting Nck may serve in T cell activation, it is interesting to speculate on its potential function. First, Nck is known to bind components of the actin cytoskeleton, e.g., the Pak/Pix complex (Buday et al. 2002). Additionally, cytoskeletal rearrangements are required for clustering of the TCR and driving the formation of an immune synapse (Campi et al. 2005; Yokosuka et al. 2005). At this point, however, the purpose of this structure is unclear, as the T cell is already well on its way to being activated by the time a synapse has been formed (Varma et al. 2006). It is also worth mentioning that phosphorylation of the ITAM within CD3ϵ disrupts Nck binding (Gil et al. 2002, 2005; Kesti et al. 2007), thus providing an additional temporal aspect to this function. Indeed, this may be an important sensor of signal strength. As our own work and that of a recent study have shown, there are clear distinctions to be made between the signals generated by the TCR leading to either proliferation or apoptosis (B. Schraven; unpublished observations; Daniels et al. 2006). However, whether Nck plays any role in this process has yet to be demonstrated.

6 Qualitative Modeling

To better understand the complexity and interactions of the many proteins involved in transducing signals from the TCR to the level of the transcription factors, we undertook the task of constructing a logical model of T cell activation. We began by considering only the minimal number of receptors necessary to activate a T cell: the TCR/CD3/ζ complex, the coreceptor CD4/CD8, and the costimulatory molecule CD28. By evaluating the published literature, we then constructed our network model step by step, considering only direct interactions from molecule to molecule. Importantly, for each interaction we required at least two independent sources that considered data only from T cells, and one citation had to exist for data generated in primary human or mouse T cells. We considered only data generated using naïve peripheral blood T cells, disregarding much of the work on thymic development; importantly, for the molecules showing a strong alteration in thymic development, we were careful to exclude those data where it appeared that the peripheral T cells may have evolved from an aberrant population.

The T cell activation network wasconstructed using CellNetAnalyzer (CNA) software (Klamt et al. 2006). Since we built a logical model, molecules can be only on or off. The reasons for this simplification were that, first, many of the data within the literature lack quantification, and second, since the data have been generated in different systems, using different types of stimuli and different time points, it was difficult to make a meaningful comparison. One additional problem arose when we considered feedback loops, since a logical model would not allow molecules to be both simultaneously on and off. This problem was overcome by including time scales, e.g., τ=1 (early) and τ=2 (late), to segregate the early and late events within the signaling cascades. Let us consider two examples that justify this approach.

The first example involves the phosphoprotein associated with glycosphingolipid-enriched microdomains (PAG) also known as the Csk binding protein (Cbp) (Brdicka et al. 2000; Kawabuchi et al. 2000). PAG/Cbp (hereafter referred to as PAG) is a transmembrane adaptor that negatively regulates the activity of Src family kinases by recruiting their regulator, the C-terminal Src kinase (Csk), to the plasma membrane where Csk then phosphorylates the inhibitory tyrosine residue conserved within the cytoplasmic tail of the Src kinases (Okada et al. 1991). When phosphorylated, the inhibitory tyrosine can then bind intramolecularly to the SH2 domain leading to the formation of the closed/inactive conformation (Sicheri et al. 1997; Xu et al. 1997). In resting T cells, PAG is one of the most abundant phosphoproteins, and Csk can be found associated via the binding of its SH2 domain to pY^{317} within PAG (Brdicka et al. 2000; Takeuchi et al. 2000). However, upon stimulation PAG is rapidly dephosphorylated, releasing Csk and allowing activation of the Src kinases (Cary and Cooper 2000). At a later time point, PAG is rephosphorylated by the Src kinase Fyn (Yasuda et al. 2002), allowing the re-recruitment of Csk and terminating signaling. Indeed, it has been shown that Fyn-deficient mice demonstrate an enhanced activation (i.e., IL-2 production) due the lack of a negative feedback mechanism(s), and indeed PAG phosphorylation is missing in these mice (Filby et al. 2007). Since the rephosphorylation of PAG is temporally distinguishable from LAT phosphorylation, which is a marker of activation (Torgersen et al. 2001), this allows us to separate these two events into discrete steps, bringing our model closer to describing the actual biology of the system.

The second example of a feedback loop involves the protein tyrosine phosphatase Shp-1 (PTPN6). It has been proposed that one mechanism capable of discriminating between activating (agonist) and non-activating (antagonist) TCR ligands involves the activation of Shp-1 by the Src kinase Lck (Stefanová et al. 2003). Using this example, it is obvious that Lck must first become activated before it can activate Shp-1, which then, in turn, inactivates Lck at a later time point. Thus, to include this feedback loop into the model it was necessary to divide it into discrete steps for the reasons stated above. Indeed, the data presented by Stefanová et al. (2003) show that Lck is activated within 1 min of TCR crosslinking, while Shp-1 association occurs only after 20–40 min, thus justifying the segregation of these processes into distinct steps.

As published (Saez-Rodriguez et al. 2007), the network comprises 94 nodes and 123 interactions. Nodes can be switched on or off to mimic both knock-in or knock-out situations. An additional feature is the ability to specify output patterns and ask the model to identify the minimal number of molecules required [called minimal intervention sets (MISs)]; similar to a reverse engineering approach. During the validation phase of the modeling we were able to reproduce several knock-out mouse phenotypes. It was during this phase that we made an unexpected observation. When we specified the input CD28 alone [i.e., CD28 (on), TCR (off), CD4/8 (off)] we noticed that JNK would become activated (via the pathway Vav1→Rac1→MLK3→MKK4→JNK). This observation was then tested and experimentally verified. Interestingly, we could not inhibit the CD28-induced JNK activation with either PP2 (a Src kinase inhibitor) or wortmannin (a PI3K inhibitor). This led us to suspect that the kinase responsible was most likely a member of the Tec family of kinases; however, as Tec kinase inhibitors were not commercially available, we could not verify this hypothesis and decided to include the missing enzyme as "kinase *x*" to indicate that this enzyme still requires identification. We then went back to the literature to see if anyone else had previously made this observation, and indeed we found one citation supporting this observation (Gravestein et al. 1998). Recall that in building our model we required two independent sources; this citation therefore had not been included in the construction phase. Thus, even this simplistic approach to modeling large networks has demonstrated predictive power. At this point our T cell network is far from complete. There are a number of other receptors and signaling pathways that need to be included that are known to function as costimulatory molecules either enhancing or inhibiting T cell activation (or in some cases both, depending on when the stimulus is applied). Additionally, we need to better integrate the temporal aspects of these signaling processes to allow a more realistic simulation of the dynamic behavior within the network, particularly with respect to the feedback loops.

7 Quantitative Modeling

In addition to the logical model of T cell activation described in the previous section, we are also attempting to generate a quantitative model describing a small portion of this network, namely TCR to ERK activation, using splenic T cells purified from OT-I transgenic mice as a model system. (Note that the ligand for the OT-I TCR is the SIINFEKL peptide of ovalbumin presented on the murine MHC class I molecule H-$2K^b$.) We began our analysis by measuring the activation kinetics of key molecules within our network to two different stimuli, peptide–MHC streptamers (i.e., the physiological ligand) or soluble antibodies to both CD3 and CD8. This approach allowed us to trigger the same receptors in both cases. Note as well that for stimulation the molecules were biotinylated to allow crosslinking.

The results of this analysis have revealed some striking differences in how T cells behave to what one might perceive as comparable stimuli (B. Schraven, unpublished observations). First, T cells stimulated with streptamers proliferate, whereas those stimulated with soluble antibodies do not. Instead, the latter appear to be driven into apoptosis. On the molecular level, one observes that for most molecules, streptamer stimulation induces a weak but sustained signaling, while antibodies induce a strong and transient activation. Interestingly, when one looks at the early events, e.g., ZAP-70 activation, one sees that although the magnitude of the signals is different, the kinetic properties are quite similar. This appears to be true also for LAT phosphorylation, but not PLCγ1 activation (as measured by pY^{783} staining). It therefore appears that the signal bifurcates at the level of LAT. The next question is, Why? It is worth noting that antibody stimulation also induces a rapid ubiquitinylation and degradation of ZAP-70, whereas streptamers do not. However, the question of whether one is able to convert an apoptotic stimuli into a proliferative one (or vice versa) by pharmacological or molecular intervention will require further investigation. An additional and important result of this study is that the strong calcium signal induced by antibodies does not necessarily correlate with activation (i.e., proliferation) of the cell.

Another extremely interesting observation that came out of this work is that when one analyzes the localization of ERK within the cell following stimulation by confocal microscopy one sees a rapid co-capping of phospho-ERK with the TCR, whereas streptamer stimulation induces more a co-clustering than capping; at later time points, one sees that these structures no longer localize to the plasma membrane, but rather in structures below the membrane (B. Schraven, unpublished observation). These results are quite similar to those published for streptamer stimulation of OT-I thymocytes (Daniels et al. 2006). In this study, stimuli that led to proliferation also induced a sustained activation of ERK intracellularly, while apoptotic stimuli induced an activation that colocalized with the TCR. Clearly, these two studies demonstrate that it is no longer sufficient to consider only when a molecule becomes activated, but we must now also ask where in the cell is that activation event occurring.

8 Future Perspectives

This now brings us to one of the more challenging aspects of modeling: getting the data. In addition to quantifying the number of molecules present in a cell, we also require high-throughput methods to rapidly generate quantitative data. Establishing these methods, however, will be challenging. Western blotting is by far the most established method for studying signal transduction, but we are limited in the number of samples that can be loaded onto the gel, and detection with film is often far from quantitative or, if it is quantitative, it is so only within a limited range. This has been drastically improved with quantitative imaging systems that now allow real-time acquisition of the signal; these systems, on the other hand, tend to be slower in sample acquisition.

The recent appearance of several bead-based systems that allow for a rapid analysis using relatively small amounts of material may help to change this. An additional advantage of these systems is the use of fluorescent dyes that allow for accurate quantification. One limitation that applies to all methods that utilize cell lysates is that one must never forget that we are averaging the response from millions of cells. Are the results a sign of a few cells responding strongly or all cells responding weakly? An additional problem is that while many antibodies have been optimized for blotting, it still remains to be seen whether these reagents will work equally well when working with proteins in their native conformation.

While intracellular staining combined with flow cytometry overcomes the problem of averaging cell responses, it still averages the signal over one cell. The number of antibodies available also limits this technique and, as not all antibodies work with the same permeabilization conditions, certain analyses may be impossible to obtain. Additional controls for specificity are required, as phospho-sites may be highly conserved within a protein family. All things considered, however, intracellular flow appears to be an excellent and developing technique, particularly as it yields statistically relevant data. An additional advantage is that many of the conditions and reagents used for flow cytometry are also applicable to microscopy, although flow cytometry tends to be much more sensitive.

Confocal fluorescence microscopy is one technique that has become essential, although whether it will ever become high throughput is open to debate. The optimization of antibodies and the number of necessary controls will certainly slow things down, at least in the near future, but perhaps automation can be utilized to speed the processes up a bit. Such issues notwithstanding, who is not overcome with awe when watching the multicolor images of signaling processes as they occur in 3D? Finally, with the advances in two-photon microscopy that now allow us to visualize cell–cell interactions as they occur in vivo, perhaps one day we will be able to utilize this technique to visualize signaling processes themselves.

Acknowledgements. The authors would like to thank Luca Simeoni and Tilo Beyer for helpful discussion and apologize to those colleagues whose work we have not cited particularly with respect to the complexity of TCR signaling. This work was supported in part by grants from the German Research Society (DFG FOR521), the German Ministry of Education and Research (BMBF FORSYS program) and the Sachsen-Anhalt Ministry of Education (Research Focus Dynamic Systems).

References

Brdicka T, Pavlistová D, Leo A, Bruyns E, Korínek V, Angelisová P, Scherer J, Shevchenko A, Hilgert I, Cerný J, Drbal K, Kuramitsu Y, Kornacker B, Horejsí V, Schraven B (2000) Phosphoprotein associated with glycosphingolipid-enriched microdomains (PAG), a novel ubiquitously expressed transmembrane adaptor protein, binds the protein tyrosine kinase csk and is involved in regulation of T cell activation. J Exp Med 191:1591–1604

Buday L, Wunderlich L, Tamas P (2002) The Nck family of adapter proteins: regulators of actin cytoskeleton. Cell Signal 14:723–731

Bunnell SC, Singer AL, Hong DI, Jacque BH, Jordan MS, Seminario MC, Barr VA, Koretzky GA, Samelson LE (2006) Persistence of cooperatively stabilized signaling clusters drives T cell activation. Mol Cell Biol 26:7155–7166

Campi G, Varma R, Dustin ML (2005) Actin and agonist MHC-peptide complex-dependent T cell receptor microclusters as scaffolds for signaling. J Exp Med 202:1031–1036

Cary LA, Cooper JA (2000) Molecular switches in lipid rafts. Nature 404:945–947

Chang JT, Palanivel VR, Kinjyo I, Schambach F, Intlekofer AM, Banerjee A, Longworth SA, Vinup KE, Mrass P, Oliaro J, Killeen N, Orange JS, Russell SM, Weninger W, Reiner SL (2007) Asymmetric T lymphocyte division in the initiation of adaptive immune responses. Science 315:1687–1691

Daniels MA, Teixeiro E, Gill J, Hausmann B, Roubaty D, Holmberg K, Werlen G, Hollander GA, Gascoigne NR, Palmer E (2006) Thymic selection threshold defined by compartmentalization of Ras/MAPK signalling. Nature 444:724–729

Fernandez-Miguel G, Alarcon B, Iglesias A, Bluethmann H, Alvarez-Mon M, Sanz E, de la Hera A (1999) Multivalent structure of an alphabetaT cell receptor. Proc Natl Acad Sci U S A 96:1547–1552

Filby A, Seddon B, Veldhoen M, Sanchez-Morgado J, Smida M, Lindquist JA, Schraven B, Zamoyska R (2007) Fyn regulates the duration of TCR engagement needed for commitment to effector function. J Immunol 179:4635–4644

Gil D, Schamel WW, Montoya M, Sanchez-Madrid F, Alarcon B (2002) Recruitment of Nck by CD3 epsilon reveals a ligand-induced conformational change essential for T cell receptor signaling and synapse formation. Cell 109:901–912

Gil D, Schrum AG, Alarcon B, Palmer E (2005) T cell receptor engagement by peptide-MHC ligands induces a conformational change in the CD3 complex of thymocytes. J Exp Med 201:517–522

Gravestein LA, Amsen D, Boes M, Calvo CR, Kruisbeek AM, Borst J (1998) The TNF receptor family member CD27 signals to Jun N-terminal kinase via Traf-2. Eur J Immunol 28:2208–2216

Kawabuchi M, Satomi Y, Takao T, Shimonishi Y, Nada S, Nagai K, Tarakhovsky A, Okada M (2000) Transmembrane phosphoprotein Cbp regulates the activities of Src-family tyrosine kinases. Nature 404:999–1003

Kesti T, Ruppelt A, Wang JH, Liss M, Wagner R, Taskén K, Saksela K (2007) Reciprocal regulation of SH3 and SH2 domain binding via tyrosine phosphorylation of a common site in CD3epsilon. J Immunol 179:878–885

Klamt S, Saez-Rodriguez J, Lindquist JA, Simeoni L, Gilles ED (2006) A methodology for the structural and functional analysis of signaling and regulatory networks. BMC Bioinformatics 7:56

Minguet S, Swamy M, Alarcon B, Luescher IF, Schamel WW (2007) Full activation of the T cell receptor requires both clustering and conformational changes at CD3. Immunity 26:43–54

Okada M, Nada S, Yamanashi Y, Yamamoto T, Nakagawa H (1991) CSK: a protein-tyrosine kinase involved in regulation of src family kinases. J Biol Chem 266:24249–24252

Osman N, Turner H, Lucas S, Reif K, Cantrell DA (1996) The protein interactions of the immunoglobulin receptor family tyrosine-based activation motifs present in the T cell receptor zeta subunits and the CD3 gamma, delta and epsilon chains. Eur J Immunol 26:1063–1068

Saez-Rodriguez J, Simeoni L, Lindquist JA, Hemenway R, Bommhardt U, Arndt B, Haus UU, Weismantel R, Gilles ED, Klamt S, Schraven B (2007) A logical model provides insights into T cell receptor signaling. PLoS Comput Biol 3:e163

San Jose E, Sahuquillo AG, Bragado R, Alarcon B (1998) Assembly of the TCR/CD3 complex: CD3 epsilon/delta and CD3 epsilon/gamma dimers associate indistinctly with both TCR alpha and TCR beta chains. Evidence for a double TCR heterodimer model. Eur J Immunol 28:12–21

Sicheri F, Moarefi I, Kuriyan J (1997) Crystal structure of the Src family tyrosine kinase Hck. Nature 385:602–609

Stefanová I, Hemmer B, Vergelli M, Martin R, Biddison WE, Germain RN (2003) TCR ligand discrimination is enforced by competing ERK positive and SHP-1 negative feedback pathways. Nat Immunol 4:248–254

Takeuchi S, Takayama Y, Ogawa A, Tamura K, Okada M (2000) Transmembrane phosphoprotein Cbp positively regulates the activity of the carboxyl-terminal Src kinase, Csk. J Biol Chem 275:29183–29186

Torgersen KM, Vang T, Abrahamsen H, Yaqub S, Horejsi V, Schraven B, Rolstad B, Mustelin T, Tasken K (2001) Release from tonic inhibition of T cell activation through transient displacement of C-terminal Src kinase (Csk) from lipid rafts. J Biol Chem 276:29313–29318

Varma R, Campi G, Yokosuka T, Saito T, Dustin ML (2006) T cell receptor-proximal signals are sustained in peripheral microclusters and terminated in the central supramolecular activation cluster. Immunity 25:117–127

Xu W, Harrison SC, Eck MJ (1997) Three-dimensional structure of the tyrosine kinase c-Src. Nature 385:595–602

Yasuda K, Nagafuku M, Shima T, Okada M, Yagi T, Yamada T, Minaki Y, Kato A, Tani-Ichi S, Hamaoka T, Kosugi A (2002) Cutting edge: Fyn is essential for tyrosine phosphorylation of Csk-binding protein/phosphoprotein associated with glycolipid-enriched microdomains in lipid rafts in resting T cells. J Immunol 169:2813–2817

Yokosuka T, Sakata-Sogawa K, Kobayashi W, Hiroshima M, Hashimoto-Tane A, Tokunaga M, Dustin ML, Saito T (2005) Newly generated T cell receptor microclusters initiate and sustain T cell activation by recruitment of Zap70 and SLP-76. Nat Immunol 6:1253–1262

Ernst Schering Foundation Symposium Proceedings, Vol. 3, pp. 63–82
DOI 10.1007/2789_2007_071

Published Online: 29 February 2008

Solving the IRAK-4 Enigma: Application of Kinase-Dead Knock-In Mice

M. Koziczak-Holbro(✉), C. Joyce, A. Glück, B. Kinzel, M. Müller, H. Gram

Novartis Institutes for BioMedical Research, Postfach 4002, Basel, Switzerland
email: *magdalena.koziczak@novartis.com*

Abstract. Interleukin-1 receptor-associated kinase (IRAK-4) is an essential component of the signal transduction complex downstream of the interleukin (IL)-1- and Toll-like receptors. Though regarded as the first kinase in the signaling cascade, the role of IRAK-4 kinase activity versus its scaffold function has been controversial. In order to investigate the role of IRAK-4 kinase function in vivo, we generated "knock-in" mice where the wild-type IRAK-4 gene is replaced with a mutant gene encoding kinase-deficient IRAK-4 protein (IRAK-4 KD). IRAK-4 kinase is rendered inactive by mutating the conserved lysine residues in the ATP pocket essential for coordinating ATP. Analyses of embryonic fibroblasts and macrophages obtained from IRAK-4 KD mice demonstrated lack of cellular responsiveness to stimulation with IL-1β or Toll-like receptor 4 (TLR4) and TLR7 agonists. IRAK-4 KD cells were severely impaired in NF-κB, JNK, and p38 activation in response to IL-1β or TLR7 ligand. In addition, activation of JNK and p38 was affected in lipopolysaccharide (LPS)-stimulated IRAK-4 KD macrophages. As a consequence, IL-1 receptor/TLR4/TLR7-mediated production of cytokines and chemokines was largely absent in these cells. Additionally, microarray analysis identified IL-1β response genes and revealed that the induction of IL-1β-responsive mRNAs is largely ablated in IRAK-4 KD cells. In summary, our results suggest that IRAK-4 kinase activity plays a critical role in IL-1R-, TLR4-, and TLR7-mediated induction of inflammatory responses.

1 Introduction

Interleukin (IL)-1 receptor (IL-1R), IL-18 receptor (IL-18R), and Toll-like receptors (TLRs) are important mediators of innate immune responses (Takeda and Akira 2005). The signaling cascades initiated by these receptors are involved in host defense mechanisms, fever induction, acute and chronic inflammation, obesity, and immune modulation. IL-1β-mediated activation of IL-1R leads to induction of cellular signaling pathways involving a cascade of adaptor molecules, kinases, and transcription factors. Although IL-1R and TLRs exhibit structurally distinct extracellular domains, they share an intracellular Toll/IL-1R (TIR) homology domain essential for interactions with downstream signaling components (Means et al. 2000). Stimulation of the IL-1R triggers engagement of the IL-1R accessory protein (IL-1RacP), followed by binding of the intracellular adaptor protein myeloid differentiation factor 88 (MyD88) via interactions of TIR domains (Fig. 1). This leads to recruitment of IL-1R-associated kinases (IRAKs), IRAK-4, IRAK-1, and tumor necrosis factor associated factor 6 (TRAF6), to the receptor complex (Martin and Wesche 2002; Janssens and Beyaert 2003). As a consequence, IRAK-1 is phosphorylated (Li et al. 2002; Kollewe et al. 2004) and later ubiquitinylated and degraded (Yamin and Miller 1997). Hyperphosphorylated IRAK-1 together with TRAF6 leaves the receptor complex and interacts with the transforming growth factor β-activated kinase 1 (TAK1) multiprotein signalosome (Jiang et al. 2002).

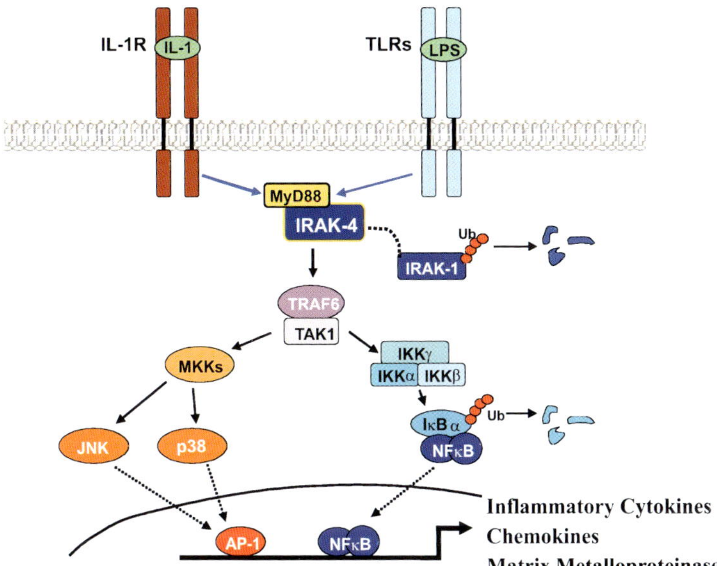

Fig. 1. A schematic representation of the current model of the IL-1β and TLR/MyD88-dependent signaling pathway. The association of IL-1 with the receptor activates a complex intracellular signaling pathway involving many adaptor proteins and kinases to induce transcription of proinflammatory genes. *IκB*, inhibitory protein B; *IKK*, inhibitory protein B kinase; *IL-1*, interleukin-1; *IL-1R*, IL-1 receptor; *IRAK*, IL-1 receptor associated kinase; *JNK*, c-Jun N-terminal kinase; *LPS*, lipopolysaccharide; *MKK*, MAPK kinase; *MyD88*, myeloid differentiation factor 88; *NF-κB*, nuclear factor-κB; *TAK1*, transforming growth factor β activated kinase 1; *TLRs*, Toll-like receptors; *TRAF6*, tumor necrosis factor-associated factor 6; *Ub*, ubiquitin

Subsequently, this new complex triggers induction of downstream signaling events, including IκB kinases (IKKs), p38, and Jun-N-terminal kinases (JNK) (Wang et al. 2001), leading to activation of transcription factors such as nuclear factor (NF)-κB and activator protein (AP)-1. These transcription factors are involved in control of expression of many proinflammatory genes encoding cytokines, chemokines, adhesion molecules, and proteolytic enzymes (Tak and Firestein 2001; Wisdom 1999).

As TLRs also signal via TIR domains, their signaling is entirely or in part dependent upon IRAK-4 (Takeda and Akira 2004).

IRAK-4 has been reported to be pivotal for IL-1R-, IL-18R-, and TLR-induced signaling (Suzuki et al. 2002b). Analysis of IRAK-4-deficient mice (IRAK-$4^{-/-}$) revealed that IRAK-4 is essential for mediating the majority of innate immune responses (Suzuki et al. 2002a). Macrophages and mouse embryonic fibroblasts (MEFs) derived from IRAK-$4^{-/-}$ mice exhibit severe defects in cellular signaling in response to IL-1β and TLR ligands, namely suppression of IκBα degradation and NF-κB activation, and inhibition of c-jun and p38 phosphorylation. Consequently, the production of proinflammatory cytokines, such as IL-6, IL-1β, and tumor necrosis factor α (TNF-α) is reduced. The animals that lack IRAK-4 also fail to produce cytokines in response to LPS challenge (Suzuki et al. 2002a). In addition, these mice are sensitive to bacterial infections. Recently, patients with recurrent infections and devoid of IRAK-4 protein due to gene mutations have been described (Medvedev et al. 2005; Picard et al. 2003). Their phenotype is reminiscent of the phenotype observed in mice devoid of TLR signaling in regard to increased susceptibility for bacterial infections. While the crucial function of IRAK-4 protein in signaling has been demonstrated, controversial data exist concerning the nature of its kinase activity. On one hand, reconstitution of mouse-derived IRAK-$4^{-/-}$ cells with kinase-inactive IRAK-4 has been insufficient to rescue IL-1β-induced activation of NF-κB and JNK, compared to overexpression of wild-type IRAK-4 (Lye et al. 2004). This implies that the kinase activity of IRAK-4 is a key requirement for the optimal transduction of IL-1β-induced signals and production of inflammatory cytokines. In contrast, similar experiments, but using human IRAK-$4^{-/-}$ cells derived from IRAK-4-deficient patients, have demonstrated that an IRAK-4 kinase-inactive mutant effectively restored IL-1β-mediated signaling, suggesting that IRAK-4 acts more as an adaptor protein rather than an essential kinase (Qin et al. 2004).

To clarify the biochemical function of IRAK-4 kinase, we have generated genetically engineered mice expressing a kinase-deficient mutant of this protein (IRAK-4 KD) (Koziczak-Holbro et al. 2007). MEFs and bone marrow-derived macrophages (BMDMs) isolated from these mice were characterized for their ability to respond to stimulation via the IL-

1R or TLRs. Our recently published analysis demonstrated that IRAK-4 kinase activity is crucial for IL-1R- and TLR- mediated signaling and expression of proinflammatory genes in these cells (Koziczak-Holbro et al. 2007). Additionally, we identified a set of genes whose expression is dependent upon the presence of IRAK-4 kinase activity. The current article summarizes our studies on IRAK-4 kinase function in inflammatory responses.

2 Results

2.1 The Kinase Activity of IRAK-4 is essential for IL-1β-Mediated Signaling Pathways

To investigate the physiological role of IRAK-4 kinase activity, "knock-in" mice were generated by replacing the wild-type (WT) gene with a gene containing a mutation in the IRAK-4 kinase domain (Fig. 2). The previously described KK213AA mutation replaces two conserved lysine residues in the ATP binding site that are essential for the hydrolysis of ATP (Li et al. 2002; Zheng et al. 1993). This mutation abolishes IRAK-4 kinase activity without affecting other protein features, namely interaction with the receptor complex via MyD88 adaptor protein (Li et al. 2002).

The knock-in approach provides a system where all signaling components, including IRAK-4 KD protein, are present at physiological levels. These mice therefore differ from the previously described IRAK-$4^{-/-}$ mice, which are completely devoid of IRAK-4 protein and do not allow investigators to differentiate between IRAK-4's function as a scaffold protein and a kinase (Suzuki et al. 2002a).

The effect of IRAK-4 kinase deficiency on the signaling pathways was studied in MEFs isolated from IRAK-4 KD mice. Upon IL-1β stimulation, the degradation of IκBα and phosphorylation of p65 NF-κB did not occur in IRAK-4 KD MEFs, but was apparent in WT MEFs (Fig. 3A). Similarly, phosphorylation of the other IL-1β-induced signaling molecules of the MAPK pathway, namely MKK4, JNK, and p38, was absent only in IRAK-4 KD MEFs. On the other hand, IL-1β-stimulated protein kinase B (PKB) and extracellular signal-regulated kinase (ERK)1/2 phosphorylation was induced equally in both WT and

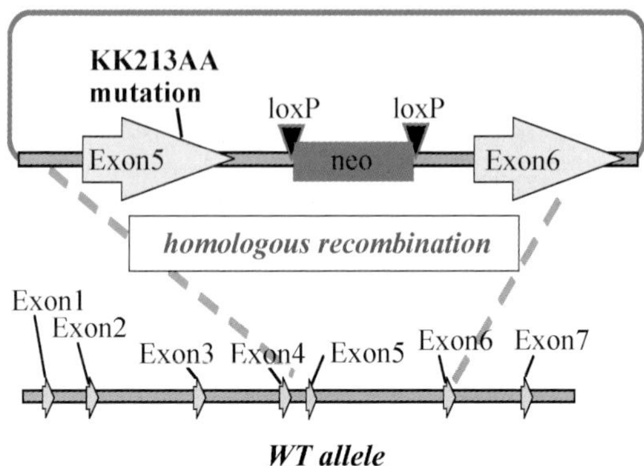

Fig. 2. Generation of IRAK-4 KD mice. To generate a targeting construct for homologous recombination in mouse embryonic stem (ES) cells, the mouse IRAK-4 genomic sequence spanning the region from intron 4 to intron 6 was amplified and subcloned into vector pRAY2. Site-directed mutagenesis was performed to change KK213–214 to AA in exon 5. The targeting plasmid was transfected into ES cells from Balb/c mice. Transfected ES cells were selected for neomycin resistance. Homologous recombination was identified by PCR and confirmed by Southern hybridization. The loxP-flanked neomycin cassette was eliminated by co-expression of Cre recombinase. The neomycin-sensitive ES cell clones were amplified and analyzed by PCR and Southern blot. ES cells with homologous recombination were injected into C57Bl/6 host blastocysts. Germline transmission in F1 heterozygous offspring was verified by PCR analysis, and F1 heterozygotes were interbred to obtain homozygous mutant mice

IRAK-4 KD cells, suggesting IRAK-4 kinase-independent regulation of these pathways downstream of IL-1R (Fig. 3B).

2.2 The Kinase Activity of IRAK-4 Is Important for IL-1β-Mediated Transcriptional Regulation

Immunoblot analysis showed that IL-1β-mediated NF-κB activation was absent in IRAK-4 KD MEFs. To further examine the consequences of disrupting this activity we compared the responsiveness of the NF-κB reporter gene to different stimuli in WT and IRAK-4 KD MEFs. The

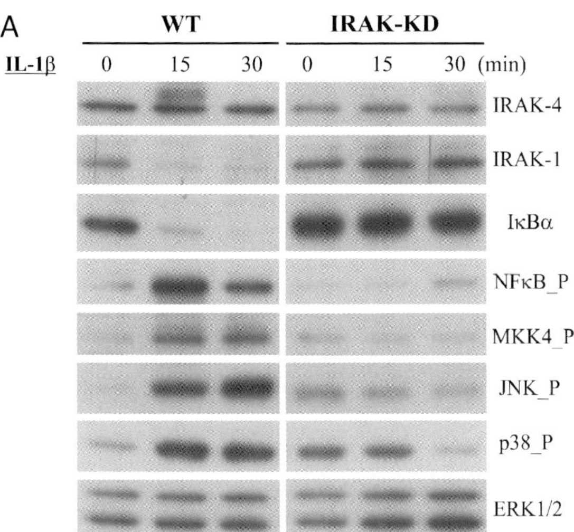

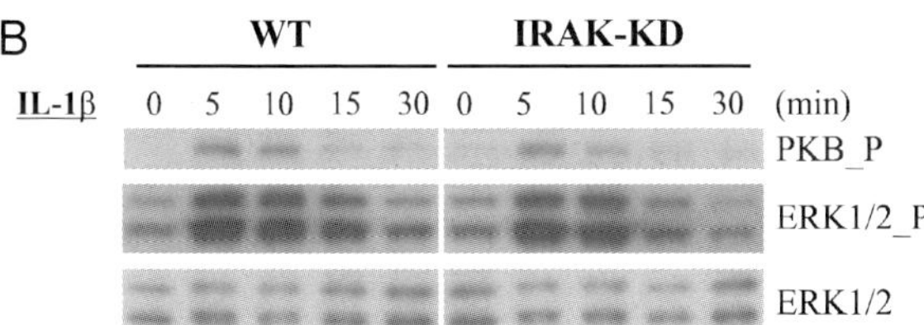

Fig. 3A, B. Effect of IRAK-4 kinase deficiency on IL-1β-mediated signaling. Wild-type (*WT*) and IRAK-4 KD MEF cells were left untreated or stimulated with IL-1β (1 ng/ml) and cell lysates were prepared after the indicated periods of time. **A** Protein levels were assayed by immunoblotting with IRAK-4, IRAK-1, and IκBα-specific antibodies. The phosphorylation status of p65 NF-κB (*NFκB_P*), MKK4 (*MKK4_P*), JNK (*JNK_P*), and p38 (*p38_P*) was evaluated by immunoblotting with phosphopeptide-specific antibodies. **B** The phosphorylation status of PKB (*PKB_P*) and ERK1/2 (*ERK1/2_P*) was also evaluated by immunoblotting with phosphopeptide-specific antibodies. Equivalent loading was confirmed by reprobing with anti-ERK1/2 antibody. The results are representative of three independent experiments

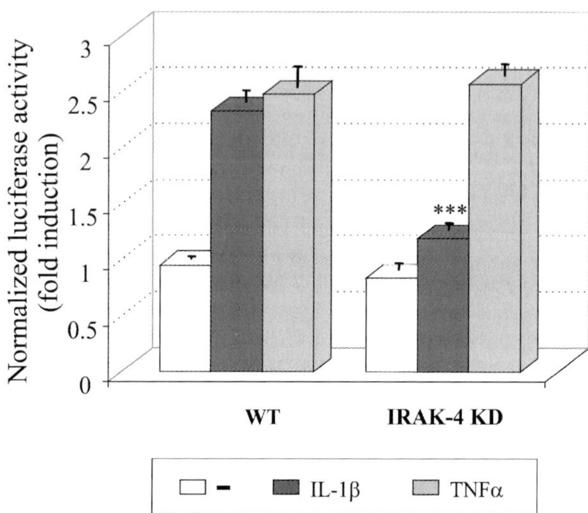

Fig. 4. Effect of IRAK-4 kinase deficiency on an NF-κB-regulated promoter. WT or IRAK-4 KD MEF cells were transfected with NF-κB-dependent luciferase reporter plasmid and control Renilla reporter plasmid. After 24 h, cells were left untreated or treated with IL-1β (1 ng/ml) or TNF-α (10 ng/ml) for 8 h and assayed for luciferase and Renilla activities. The luciferase activity was normalized with the Renilla activity, and fold activation was calculated relative to the activity of WT untreated cells (set as one-fold). The results presented are the means±SEM of two independent experiments performed in triplicate. $^{***}p < 0.001$ indicates a significant difference between WT and IRAK-4 KD samples of the same treatment

activation of the promoter by both IL-1β and TNF-α via the NF-κB element was apparent in WT cells (Fig. 4). However, the kinase deficiency of IRAK-4 prevented IL-1β-mediated, but not TNF-α-mediated, stimulation of the promoter. These data confirmed that the kinase activity of IRAK-4 is essential for NF-κB induction and that the effect we observed is specific for IL-1β-stimulated pathways.

2.3 IRAK-4 KD Affects Expression of Proinflammatory Genes

Based on the fact that IL-1β-induced transcriptional activity was disrupted in IRAK-4 KD MEFs, we anticipated that RNA expression of IL-1β-induced genes will also be affected. To identify these genes, we performed oligonucletide microarray experiment. After 1 h of IL-1β stimulation compared to no treatment, expression of 41 genes and expressed sequence tags (ESTs) was induced at least five-fold in WT cells. In contrast, in IRAK-4 KD cells 36 of these genes were not induced at all or had a minor increase in expression—less than 50% of WT induction (Fig. 5). The other 5 IL-1β-stimulated genes were affected to a lesser extent in IRAK-4 KD. Interestingly, the strongly affected genes belong to a group encoding proteins with known relevance to inflammation, including cytokines (IL-6, Lif), chemokines (Cxcl1, Cxcl2, Cxcl5, Ccl2, Ccl7, Ccl20), receptors (IFN-γR1), other signaling molecules (Gem, Rasa2, Rgs16, Map3k8, IκBα, TNFaip3, Pde4b, Plscr1), transcription factors (Atf3, Irf1, IκBζ, FosB), and mRNA stability regulators (Zfp36). Additionally, NF-κB-dependent antiapoptotic Birc3 and Btg2 were not induced in IRAK-4 KD cells, in contrast to WT cells. Moreover, expression of a further 54, mainly proinflammatory, genes was strongly elevated in WT but not in IRAK-4 KD cells after prolonged IL-1β stimulation, measured at 4 h (data not shown).

The results of microarray hybridization were validated by quantitative PCR analyses of selected genes (Koziczak-Holbro et al. 2007).

To confirm results obtained from mRNA quantifications, protein levels of IL-6, Cxcl1, and Ccl2 were measured in the supernatants of IL-1β-stimulated cell cultures (Fig. 6A). As expected, the production of these cytokines was significantly upregulated in WT MEFs. However, the kinase deficiency of IRAK-4 dramatically reduced the protein expression of these cytokines. The kinase deficiency of IRAK-4 did not affect TNF-α-activated cytokine production (Fig. 6B). Taken together, all these analyses demonstrate that the kinase activity of IRAK-4 is essential for the IL-1β-mediated production of most of the proinflammatory cytokines in MEFs.

A

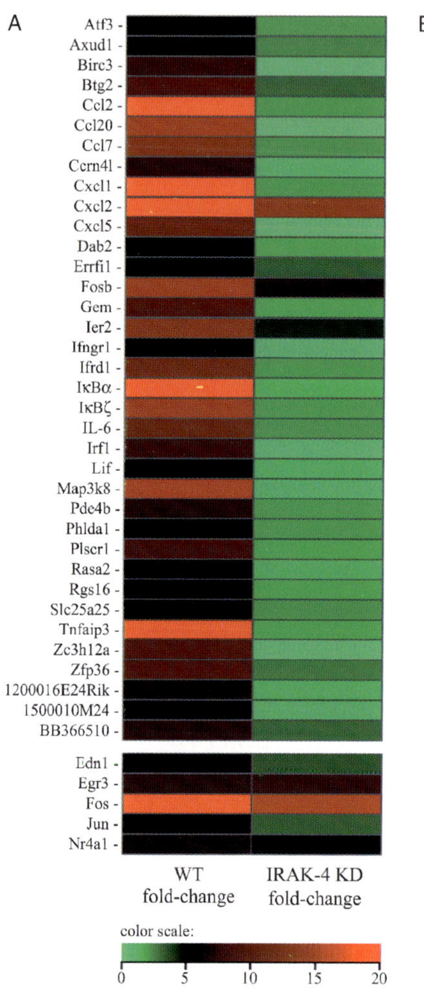

B **Affected are genes involved in inflammatory responses:**

cytokines:
IL-6
Lif

chemokines:
KC (Cxcl1)
MCP-1 (Ccl2)
MCP-3 (Ccl7)
MIP-2α (Cxcl2)
MIP-3α (Ccl20)
Cxcl5

transcription factors:
ATF3
IκBζ
Irf1
FosB

apoptosis:
Birc2
Btg2

receptors:
IFNγR1

mRNA stability regulators:
Zfp36 (TTP)

signaling molecules:
Gem (GTPase)
Pde4b (cAMP specific)
Rgs16 (GPCR signaling)
Rasa2
TNFαIP3
IκBα
TNFaip3
Plscr1

Fig. 5A, B. Microarray-based identification of IL-1β-induced and IRAK-4 kinase activity-dependent genes. WT or IRAK-4 KD MEF cells were either left untreated or stimulated with IL-1β (1 ng/ml) for 1 h. Total RNA was extracted and used for GeneChip microarray analysis. **A** Levels of gene expression are indicated as fold inductions relative to unstimulated controls. Shown are mRNAs induced more than 5-fold in WT cells. Fold regulations are presented in color code as indicated at the *bottom*: *green* for 0-fold, *black* for 5-fold, and *red* for 20-fold (or more). Genes with greater than 50% (*upper panel*) and less than 50% (*lower panel*) difference in expression between IRAK-4 KD and WT samples are presented. Gene names or identifiers corresponding to the ESTs are shown on the *left*. Data represent averages of assays done in duplicate. **B** List of affected genes separated in functional groups

2.4 IRAK-4 Kinase Activity Regulates Signaling Pathways Downstream of TLRs

Macrophages are reactive to microbial products that activate proinflammatory responses via TLR-mediated signaling. Individual TLRs transduce signals via divergent pathways: the MyD88-dependent pathway, which is highly similar to the IL-1β-mediated pathway, and/or the MyD88-independent pathway (Takeda and Akira 2005). In our studies, activation of various TLRs was tested using different specific stimuli, such as LPS (TLR4), poly I:C (TLR3), resiquimod (TLR7), and CpG (TLR9) (Fig. 7A). The BMDMs were unresponsive to CpG, but responded to the other TLR ligands. TLR3 is known to transmit signals via a MyD88/IRAK-4-independent pathway; therefore, both WT and IRAK-4 KD cells showed a normal poly I:C-induced IκB degradation and JNK phosphorylation. TLR4 can use the MyD88/IRAK-4 pathway in addition to the Trif-pathway, which is not dependent on IRAKs, while TLR7 signals only via a MyD88-dependent pathway (Takeda and Akira 2005). Both LPS- and resiquimod-stimulated IRAK-4 KD macrophages revealed disrupted activation of IκB and JNK pathways. However, the effects on TLR4-mediated signaling were not so apparent as on TLR7-mediated pathways, which is presumably due to the additional involvement of the MyD88-independent Trif pathway downstream of TLR4. As IRAK-1/4-mediated signaling is confined to the

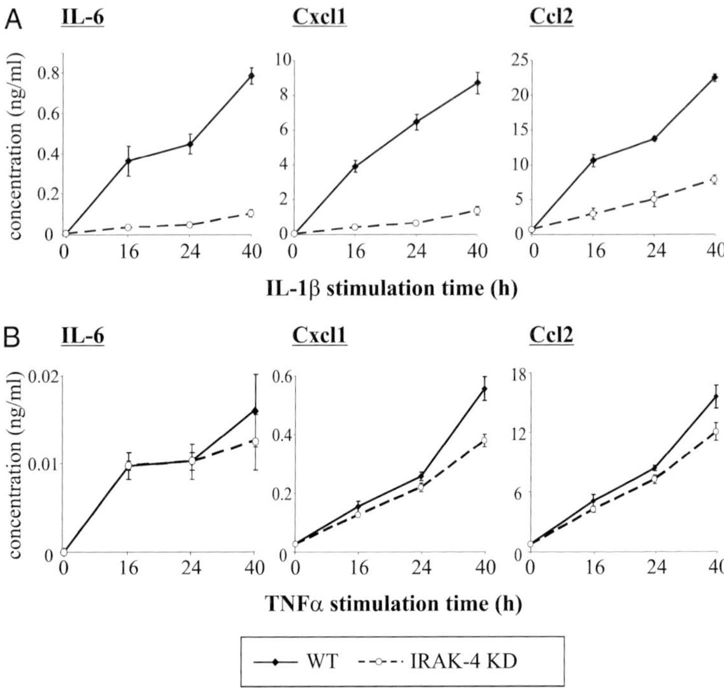

Fig. 6A, B. Effect of IRAK-4 kinase deficiency on IL-1β-induced cytokine and chemokine production. WT and IRAK-4 KD MEFs were treated with IL-1β (1 ng/ml) (**A**) or TNFα (10 ng/ml) (**B**) for the indicated times. The concentration of IL-6, Cxcl1, and Ccl2 in the culture supernatants was measured by enzyme-linked immunosorbent assay (ELISA). The results presented are the means ± SEM of two independent experiments performed in duplicate

MyD88-dependent pathway, we assessed the role of IRAK-4 kinase activity in TLR-mediated pathways by studying TLR7-induced signaling. Stimulation of TLR7 with resiquimod in WT cells led to activation of NF-κB and MAPK pathways, reflected by a decrease in levels of IRAK-1 and IκBα, and an increase in JNK, p38, and ERK1/2 phosphorylation (Fig. 7B). Conversely, these effects were significantly abolished in IRAK-4 KD BMDMs. These findings indicate that IRAK-4 kinase activity is necessary for TLR-mediated activation of MyD88-dependent

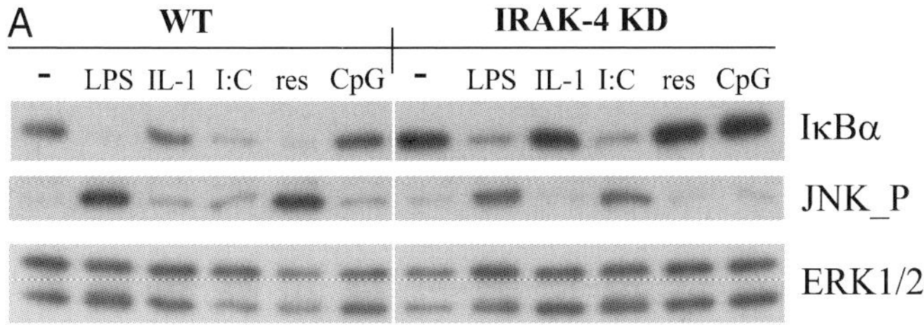

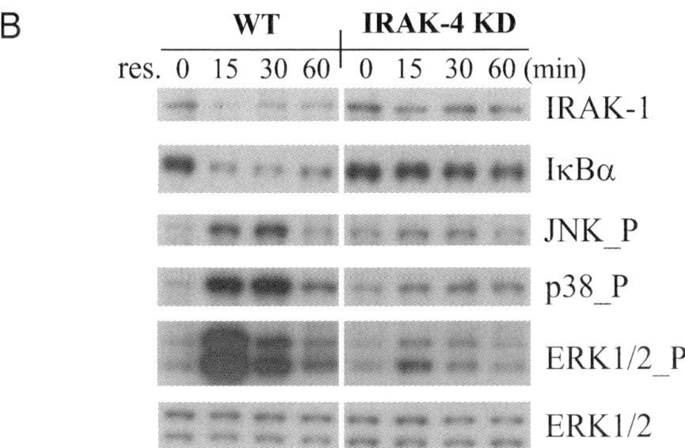

Fig. 7A, B. Effect of IRAK-4 kinase deficiency on TLR-induced signaling pathways in BMDMs. WT and IRAK-4 KD BMDMs were treated with LPS (1 μg/ml), IL-1β (IL-1; 5 ng/ml), poly I:C (25 μg/ml), resiquimod (*res.*; 0.5 μg/ml), and CpG (5 μg/ml) for 30 min (**A**) or only resiquimod (*res.*; 0.5 μg/ml) for the indicated times (**B**), and cell lysates were prepared. Protein levels were assayed by immunoblotting, as described in Fig. 3

pathways, similar to what has been observed for IL-1β-mediated signal transduction. In addition, IRAK-4 kinase activity downstream of TLR7 is required for ERK1/2 activation in BMDMs.

Since activated macrophages are directly involved in processes of inflammation and produce most of the proinflammatory mediators, we examined how kinase deficiency is affecting TLR-mediated cytokine production. While WT cells produced relatively high levels of cytokines and chemokines upon LPS or resiquimod stimulation, IRAK-4 KD cells

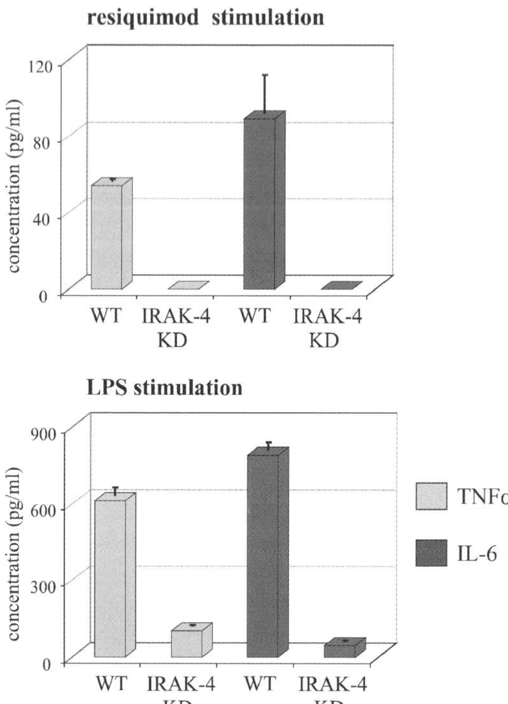

Fig. 8. Effect of IRAK-4 kinase deficiency on LPS- or resiquimod-induced cytokine production. WT and IRAK-4 KD BMDMs were treated with LPS (1 μg/ml) or resiquimod (0.5 μg/ml) for 24 h. The concentration of TNFα and IL-6 in the culture supernatants were measured by ELISA. The results presented are the means ± SEM of an experiment performed in duplicate

were unresponsive to the stimulant (Fig. 8). These data indicate that IRAK-4 kinase activity is indeed important in control of expression of proinflammatory mediators downstream of TLR7.

To examine the effects of IRAK-4 kinase deficiency on cytokine production in an in vivo situation, WT and IRAK-4 KD mice were injected intraperitoneally with LPS, the TLR4 ligand. High concentrations of TNF-α and IL-6 were detected in sera of WT animals 2 h after LPS injection (Fig. 9). In contrast, mice lacking IRAK-4 activity were resistant

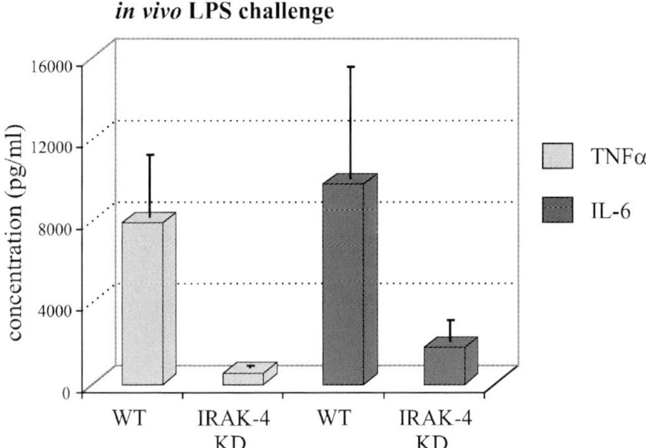

Fig. 9. In vivo LPS challenge. WT and IRAK-4 KD mice (4 mice per group) were injected intraperitoneally with LPS (20 mg/kg). Serum concentration of TNF-α and IL-6 at 2 h was measured by ELISA

to LPS challenge and showed significantly reduced cytokine response. These data showed that IRAK-4 kinase activity is critical for maintaining normal proinflammatory reactions in vivo.

3 Conclusions

Our analysis of different cell types derived from IRAK-4 KD mice clearly showed that the kinase activity of IRAK-4 is critical for IL-1R/TLR-mediated MyD88-dependent activities. Namely, IRAK-4 kinase is essential for optimal signaling activation, induction of transcription, and (as a consequence) production of proinflammatory cytokines and chemokines. Not surprisingly, the 90 genes identified by microarray analysis as strongly dependent on IRAK-4 kinase-mediated activity in MEF cells belong to the chemokines, cytokines, receptors, signaling molecules, and transcription factors known to be involved in inflammatory responses. Importantly, the fact that IRAK-4 KD animals show reduced cytokine production upon LPS challenge confirmed that IRAK-4 kinase-dependent regulation plays a significant role in vivo.

Our observation that IRAK-1 does not undergo phosphorylation and degradation in IRAK-4 KD cells stimulated with IL-1β supports the view of IRAK-4 as the IL-1/TLR first proximal kinase, suggested by previous in vitro studies with recombinant enzymes and studies of the IL-1R complex. In these studies, IRAK-1 has been proposed as a substrate of IRAK-4 kinase activity. Two crucial sites (Thr387 and Ser376) in the activation loop of IRAK-1 have been identified as a potential target for IRAK-4 kinase activity in peptide phosphorylation assays (Kollewe et al. 2004; Li et al. 2002). Additional insights into IRAK-4-mediated regulation of IRAK-1 have been gained from studies of co-expression of MyD88s, an alternatively spliced variant of MyD88. MyD88s prevents recruitment of IRAK-4 into the receptor complex, and thus inhibits IRAK-4-mediated IRAK-1 phosphorylation (Burns et al. 2003). Moreover, the recently published structure of the human IRAK-4 kinase domain revealed that IRAK-4 contains characteristic structural features of an active serine/threonine kinase (Wang et al. 2006). Also, work by Cheng et al. (2007) suggests autocatalytic phosphorylation of IRAK-4, which strongly supports the concept of IRAK-4 as the first proximal kinase in the IL-1R complex.

In our studies, we characterized in detail the IRAK-4 kinase-regulated signaling relevant for MyD88-dependent pathways. We observed complete absence of activation of the NF-κB, p38, and JNK pathways in IRAK-4 KD cells in response to stimulation with IL-1β or resiquimod, suggesting that the presence of IRAK-4 protein without a kinase activity is insufficient to relay signals from the IL-1R or TLR7. LPS-mediated activation of the NF-κB pathway was unaffected in IRAK-4 KD cells, most probably due to signaling via the MyD88-independent pathway downstream of TLR4. We also observed IRAK-4 kinase activity-dependent stimulation of ERK1/2 via TLR7 in BMDMs. This is in contrast to IRAK-4 KD MEFs, where the activation of the ERK1/2 pathway downstream of the IL-1R, previously described as MyD88-independent (Davis et al. 2006; Huang et al. 2004), was not attenuated. This observation may suggest that a receptor-specific and/or cell type-specific divergence of ERK1/2 activation exists.

The results we present here are in sharp contrast to conclusions drawn from experiments with human IRAK-4-deficient fibroblasts (Qin et al. 2004). While we believe that these discrepancies might be due to the experimental approach, we cannot formally exclude that species differences between human and mouse IRAK-4-mediated signaling exist. However, such a species disparity appears unlikely, given the high conservation throughout phylogeny of this pathway. It also appears that IRAK-4 function is not cell type-dependent as suggested by previous studies (Lye et al. 2004; Qin et al. 2004). We observed comparable results in MEF cells as well as in adult BMDMs in respect to IRAK-4 involvement in signaling and control of gene expression.

Importantly, our published data show that there is no difference in signaling pathways, as well as in mRNA expression and cytokine production, between resiquimod-induced IRAK-$4^{-/-}$ and IRAK-4 KD macrophages (Koziczak-Holbro et al. 2007). The fact that deletion of the whole protein or just elimination of its kinase features results in the blocking of pathway activation supports our notion that the kinase activity is crucial for the function of the IRAK-4 protein.

Additionally, two other groups recently used a similar knock-in approach to generate IRAK-4 kinase-deficient mice (Kawagoe et al. 2007; Kim et al. 2007). They both showed that the IRAK-4 kinase-inactive mice were completely resistant to LPS-induced shock and revealed reduced TLRs-mediated inflammatory responses in cellular assays, data which confirm our results.

In summary, the riveting discovery that IRAK-4 kinase activity has a crucial role in proinflammatory responses is not only important for understanding the basic mechanisms of IL-1/TLR-mediated gene activation, but may also have implications for the development of antiinflammatory drugs. As a next step, it would be appealing to assess the importance of IRAK-4 kinase activity in animal models of inflammatory diseases. Interestingly, it has been proposed that IL-1β and TNF-α have distinct inflammatory and destructive roles in autoimmune disease, e.g., rheumatoid arthritis (RA) (van den Berg 2001). Many animal models suggest that IL-1β has a much more potent function in the destruction of articular cartilage than TNF-α, which would have more of a predominant role in the inflammatory process. Hence inhibiting the IL-1β pathway, rather than the TNF-α pathway, could be more advantageous for the prevention of joint destruction. The benefits of interfering with the activity of IL-1β using the recombinant IL-1R antagonist anakinra have also been shown in clinical trials (Cohen 2004). The kinase activity of

IRAK-4 is thus a very attractive therapeutic target for many inflammatory diseases, in particular RA, and represents an exciting novel strategy with the potential of complementing existing therapies to improve patient recovery.

References

Burns K, Janssens S, Brissoni B, Olivos N, Beyaert R, Tschopp J (2003) Inhibition of interleukin 1 receptor/Toll-like receptor signaling through the alternatively spliced, short form of MyD88 is due to its failure to recruit IRAK-4. J Exp Med 197:263–268

Cheng H, Addona T, Keshishian H, Dahlstrand E, Lu C, Dorsch M, Li Z, Wang A, Ocain TD, Li P, Parsons TF, Jaffee B, Xu Y (2007) Regulation of IRAK-4 kinase activity via autophosphorylation within its activation loop. Biochem Biophys Res Commun 352:609–616

Cohen SB (2004) The use of anakinra an interleukin-1 receptor antagonist in the treatment of rheumatoid arthritis. Rheum Dis Clin North Am 30:365–80 vii

Davis CN, Tabarean I, Gaidarova S, Behrens MM, Bartfai T (2006) IL-1beta induces a MyD88-dependent and ceramide-mediated activation of Src in anterior hypothalamic neurons. J Neurochem 98:1379–1389

Huang Q, Yang J, Lin Y, Walker C, Cheng J, Liu ZG, Su B (2004) Differential regulation of interleukin 1 receptor and Toll-like receptor signaling by MEKK3. Nat Immunol 5:98–103

Janssens S, Beyaert R (2003) Functional diversity and regulation of different interleukin-1 receptor-associated kinase (IRAK) family members. Mol Cell 11:293–302

Jiang Z, Ninomiya-Tsuji J, Qian Y, Matsumoto K, Li X (2002) Interleukin-1 (IL-1) receptor-associated kinase-dependent IL-1-induced signaling complexes phosphorylate TAK1 and TAB2 at the plasma membrane and activate TAK1 in the cytosol. Mol Cell Biol 22:7158–7167

Kawagoe T, Sato S, Jung A, Yamamoto M, Matsui K, Kato H, Uematsu S, Takeuchi O, Akira S (2007) Essential role of IRAK-4 protein and its kinase activity in Toll-like receptor-mediated immune responses but not in TCR signaling. J Exp Med 204:1013–1024

Kim TW, Staschke K, Bulek K, Yao J, Peters K, Oh KH, Vandenburg Y, Xiao H, Qian W, Hamilton T, Min B, Sen G, Gilmour R, Li X (2007) A critical role for IRAK4 kinase activity in Toll-like receptor-mediated innate immunity. J Exp Med 204:1025–1036

Kollewe C, Mackensen AC, Neumann D, Knop J, Cao P, Li S, Wesche H, Martin MU (2004) Sequential autophosphorylation steps in the interleukin-1 receptor-associated kinase-1 regulate its availability as an adapter in interleukin-1 signaling. J Biol Chem 279:5227–5236

Koziczak-Holbro M, Joyce C, Glück A, Kinzel B, Müller M, Tschopp C, Mathison JC, Davis CN, Gram H (2007) IRAK-4 kinase activity is required for interleukin-1 (IL-1) receptor- and toll-like receptor 7-mediated signaling and gene expression. J Biol Chem 282:13552–13560

Li S, Strelow A, Fontana EJ, Wesche H (2002) IRAK-4: a novel member of the IRAK family with the properties of an IRAK-kinase. Proc Natl Acad Sci USA 99:5567–5572

Lye E, Mirtsos C, Suzuki N, Suzuki S, Yeh WC (2004) The role of interleukin 1 receptor-associated kinase-4 (IRAK-4) kinase activity in IRAK-4-mediated signaling. J Biol Chem 279:40653–40658

Martin MU, Wesche H (2002) Summary and comparison of the signaling mechanisms of the Toll/interleukin-1 receptor family. Biochim Biophys Acta 1592:265–280

Means TK, Golenbock DT, Fenton MJ (2000) The biology of Toll-like receptors. Cytokine Growth Factor Rev 11:219–232

Medvedev AE, Thomas K, Awomoyi A, Kuhns DB, Gallin JI, Li X, Vogel SN (2005) Cutting edge: expression of IL-1 receptor-associated kinase-4 (IRAK-4) proteins with mutations identified in a patient with recurrent bacterial infections alters normal IRAK-4 interaction with components of the IL-1 receptor complex. J Immunol 174:6587–6591

Picard C, Puel A, Bonnet M, Ku CL, Bustamante J, Yang K, Soudais C, Dupuis S, Feinberg J, Fieschi C, Elbim C, Hitchcock R, Lammas D, Davies G, Al-Ghonaium A, Al-Rayes H, Al-Jumaah S, Al-Hajjar S, Al-Mohsen IZ, Frayha HH, Rucker R, Hawn TR, Aderem A, Tufenkeji H, Haraguchi S, Day NK, Good RA, Gougerot-Pocidalo MA, Ozinsky A, Casanova JL (2003) Pyogenic bacterial infections in humans with IRAK-4 deficiency. Science 299:2076–2079

Qin J, Jiang Z, Qian Y, Casanova JL, Li X (2004) IRAK4 kinase activity is redundant for interleukin-1 (IL-1) receptor-associated kinase phosphorylation and IL-1 responsiveness. J Biol Chem 279:26748–26753

Suzuki N, Suzuki S, Duncan GS, Millar DG, Wada T, Mirtsos C, Takada H, Wakeham A, Itie A, Li S, Penninger JM, Wesche H, Ohashi PS, Mak TW, Yeh WC (2002a) Severe impairment of interleukin-1 and Toll-like receptor signalling in mice lacking IRAK-4. Nature 416:750–756

Suzuki N, Suzuki S, Yeh WC (2002b) IRAK-4 as the central TIR signaling mediator in innate immunity. Trends Immunol 23:503–506

Tak PP, Firestein GS (2001) NF-kappaB: a key role in inflammatory diseases. J Clin Invest 107:7–11

Takeda K, Akira S (2004) TLR signaling pathways. Semin Immunol 16:3–9

Takeda K, Akira S (2005) Toll-like receptors in innate immunity. Int Immunol 17:1–14

van den Berg WB (2001) Uncoupling of inflammatory and destructive mechanisms in arthritis. Semin Arthritis Rheum 30:7–16

Wang C, Deng L, Hong M, Akkaraju GR, Inoue J, Chen ZJ (2001) TAK1 is a ubiquitin-dependent kinase of MKK and IKK. Nature 412:346–351

Wang Z, Liu J, Sudom A, Ayres M, Li S, Wesche H, Powers JP, Walker NP (2006) Crystal structures of IRAK-4 kinase in complex with inhibitors: a serine/threonine kinase with tyrosine as a gatekeeper. Structure 14:1835–1844

Wisdom R (1999) AP-1: one switch for many signals. Exp Cell Res 253:180–185

Yamin TT, Miller DK (1997) The interleukin-1 receptor-associated kinase is degraded by proteasomes following its phosphorylation. J Biol Chem 272:21540–21547

Zheng J, Knighton DR, ten Eyck LF, Karlsson R, Xuong N, Taylor SS, Sowadski JM (1993) Crystal structure of the catalytic subunit of cAMP-dependent protein kinase complexed with MgATP and peptide inhibitor. Biochemistry 32:2154–2161

Ernst Schering Foundation Symposium Proceedings, Vol. 3, pp. 83–95
DOI 10.1007/2789_2007_072

Published Online: 29 February 2008

Sensing, Presenting, and Regulating PAMPS

J.L. de Diego(✉), G. Gerold, A. Zychlinsky

Department for Cellular Microbiology, Max Planck Institute for Infection Biology, 10117 Berlin, Germany
email: *diego@mpiib-berlin.mpg.de*

Abstract. Recognition of microbial infection and initiation of immune responses are controlled by multiple mechanisms. Toll-like receptors (TLRs) are key components of the innate immune system that detect microbial infection. TLR activation helps to eliminate the invading pathogens, coordinate systemic defenses, and initiate adaptive immune responses. Despite progress elucidating the TLR signaling aspects and the physiological relevance of TLRs in microbial infections, the molecular basis of microbial recognition by TLRs is still not fully understood. In this article we focus on the availability of microbial ligands to regulate presentation to TLRs and assist in our understanding of TLR-mediated microbial recognition.

1 Introduction

Sensing and overcoming microbial infections is essential to the survival of the host. The reliable detection of pathogens is a challenging task because of their molecular heterogeneity and rapid evolution. To solve this problem, the host has evolved a strategy based on the detection of a limited set of conserved molecular patterns that are unique to the microbial world and invariant among pathogens (Medzhitov and Janeway 1998, 2000). The pattern recognition motifs are often called pathogen-associated molecular patterns (PAMPs), although renaming them as microbe-recognition patterns or MAMPs has been proposed, since they are carried both by pathogenic and nonpathogenic organisms. PAMPs are detected by pattern recognition receptors (PRRs) that signal to the host the presence of infection. Signaling through these receptors typically leads to the production of proinflammatory cytokines and chemokines and to the expression of costimulatory molecules by antigen-presenting cells. PRR-mediated activation of antigen-presenting cells, together with the presentation of antigenic peptides, couples innate immune recognition of pathogens with the activation of adaptive immune responses.

The Toll-like receptor (TLR) family is the best-characterized class of PRRs in mammals. Thirteen TLRs have been identified from mammalian genomes, ten of which are present in humans (Roach et al. 2005). TLRs sense multiple microbial products, including lipopolysaccharide (LPS) (detected by TLR4), bacterial lipoproteins (detected by TLR2), flagellin (detected by TLR5), the unmethylated CpG DNA of bacteria and viruses (detected by TLR9), double-stranded RNA (detected by TLR3), and single-stranded viral RNA (detected by TLR7 and 8) (Akira et al. 2006). TLRs 1, 2, 4, 5, and 6 are expressed at the cell surface and recognize mainly bacterial products that are unique to prokaryotes, whereas TLRs 3, 7, 8, and 9 are specialized in viral detection and recognize nucleic structures, which are also present in the host. In this case, the accessibility to the receptors is a restriction for TLR recognition. These TLRs are localized in intracellular compartments and detect microbial nucleic acids in late endosomes–lysosomes (Diebold et al. 2004; Latz et al. 2007). Since the host's nucleic acids are not normally accessible in these compartments, they do not activate TLRs. However, in the case of deficient clearance of apoptotic cells, host DNA and RNA may become endogenous ligands for TLRs, which may lead to autoimmunity (Barrat et al. 2005; Leadbetter et al. 2002).

Although TLRs play a critical role in microbial sensing, they are not the only PRRs with this function. At the cell surface, C-type lectin-like molecules, such as the mannose receptor and the beta-glucan receptors (such as dectin-1), also participate in the recognition and uptake of microbial components (Brown 2006). Additionally, cytosolic receptors detect the presence of microbial material that gains entry into the cytoplasm. These cytosolic receptors comprise the NLR family [nucleotide-binding oligomerization domain (NOD)-like receptor family] and the caspase-recruiting domain (CARD) helicases including the retinoic-acid inducible gene I (RIG-1) and the melanoma-differentiation-associated gene 5 (MDA-5) (Lee and Kim 2007). There is growing evidence that these additional PRRs can cooperate with TLRs in the innate immune response to pathogens (Trinchieri and Sher 2007).

Using mouse models of infection, in which different TLRs or adaptor proteins involved in TLR signaling are disrupted, it has been demonstrated that TLR receptors are important in many different infections. These include tuberculosis, salmonellosis, and listeriosis among others. However, these data underscore the relevance of an adequate and timely innate immune response to bacterial pathogens (Gerold et al. 2007). Furthermore, polymorphisms in TLRs in human populations have been linked to susceptibility. Studies of a TLR2 polymorphism, for example, revealed that TLR2 is implicated in the resistance to staphylococcal infections (Lorenz et al. 2000), tuberculosis (Ogus et al. 2004), and leprosy (Krutzik et al. 2003) in humans.

Binding of PAMPs to TLRs triggers a series of events leading to antimicrobial and inflammatory responses to infection as well as to responses required for an adequate acquired immune response. The inflammatory process includes killing of microbes and marked changes in tissue physiology such as liquefaction of infected tissue to prevent microbial metastasis and the healing of tissues damaged by the host's response. If at any step inflammation is dysregulated, the inflammatory process can lead to a variety of pathological conditions, including septic shock, autoimmunity, atherosclerosis, and metabolic syndrome (Cohen 2002; Karin et al. 2006). Therefore, the inflammatory response must be tightly controlled and indeed, multiple regulatory mechanisms direct the extent and duration of TLR-induced inflammation. These include the inhibition of TLR signaling by inducible negative regulators, production of antiinflammatory cytokines, modifications of the TLR signaling complex, and chromatin remodeling of TLR-induced inflammatory genes (Foster et al. 2007; Liew et al. 2005). However, most of these regulatory mechanisms target all TLR signaling pathways and thus broadly interrupt inflammation. We reasoned, therefore, that the first line of blockage and regulation of TLR signaling exists in the body fluids and that the humoral arm of the innate immunity accounts for additional regulatory mechanisms to allow individual TLR regulation. In this article, we focus on the upstream regulatory actions of the humoral components of the innate immunity to control TLR2-induced inflammation.

2 Bacterial Lipoprotein Recognition by TLR2

TLR2 is a type I membrane proteins consisting of an extracellular leucine-rich repeat ectodomain, a single transmembrane domain, and a cytoplasmic Toll-interleukin 1 receptor (TIR) domain that resembles the cytoplasmic domain of the IL-1 receptor. The TLR2 ectodomain is responsible for the recognition of microbial products. TLR2 recognizes the anchor motif of the bacterial lipoproteins (BLP), which are ubiquitous constituents of all bacterial surfaces. This anchor motif consists of an amino-terminal cysteine [*S*-(2,3-dihydroxypropyl)-cysteine] acylated by two or three fatty chains, and it is the minimal requirement for TLR2 activation (Aliprantis et al. 1999; Brightbill et al. 1999). This lipid anchor results of the modification of a cysteine of the lipoproteins with diacylglycerol, a reaction catalyzed by the enzyme phosphatidylglycerol-prolipoprotein diacylglyceryl transferase, which is only present in prokaryotes. After the cleavage of the precursor protein, a fatty chain is attached to the amino group of the N-terminal diacylglyceryl-modified cysteine to form *N*-acyl-*S*-diacylglyceryl cysteine, the TLR2 agonist. All bacterial lipoproteins are exposed to the outer surface of the microorganism and can have very diverse functions. Experiments show that nanomolar concentrations of highly purified lipoproteins or synthetic peptides containing the lipid-modified amino-terminal cysteine [i.e: tripalmitoyl-*S*-glyceryl cysteine (Pam3Cys)] activate TLR2 (Takeuchi et al. 2000). It has been reported that TLR2 also recognizes a broad spectrum of different microbial components from viruses, bacteria, fungi, and parasites (Akira et al. 2006). Since there is no sensitive and specific biochemical method to detect BLPs, it is difficult to exclude contaminations of other biological molecules with this bacterial component.

TLR2 is distinct from other TLRs in that this receptor is activated not by homotypic interaction but by heterotypic interaction between TLR1 or TLR6 and TLR2 (West et al. 2006). In this way, the number of acyl chains in the lipopeptides is finely discriminated by TLR2/TLR1 and TLR2/TLR6. The TLR1/TLR2 heterodimer recognizes triacylated lipoproteins, whereas the TLR2/TLR6 complex senses diacylated lipoproteins (Takeuchi et al. 2001, 2002). Triacylated BLPs are carried by most prokaryotes, while diacylated lipoproteins are mainly expressed by microorganisms such as mycoplasma. In addition to the number of acyl chains, other properties such as the length or ester bonds of these acyl chains and the nature of the amino acids can also be discriminated by the different TLR2 complexes (Buwitt-Beckmann et al. 2005, 2006). Because fine differences in the structure of lipopeptides are sensed by TLR2 oligomers, gram-negative and gram-positive bacteria may take advantage of this by modifying the structure of lipopeptides to regulate the innate immune response, as was already reported for LPS in TLR4 activation (Guo et al. 1997).

3 Mechanisms Regulating Responsiveness to Bacterial Lipoproteins

TLR2 is expressed at the cell surface of many cells of the immune system including granulocytes, monocytes, myeloid dendritic, and T and B cells. Effects on TLR2 expression can lead to changes in the immune function of those cells to BLPs. Expression of TLR2 may be regulated at many levels, including at the stage of posttranslational mechanisms. The chaperone Gp96 is required for functional cell-surface expression of TLR1, TLR2, and TLR4 (Randow and Seed 2001). Gp96 associates with TLRs in the endoplasmic reticulum. In the absence of Gp96, TLRs are not functional because they are largely retained in the endoplasmic reticulum and fail to mediate responses such as TLR2-induced toxicity and host resistance by the gram-positive bacteria *Listeria monocytogenes* (Yang et al. 2007).

Coreceptors and accessory molecules have already been reported to have a regulatory role in TLR2 responses. An approach using forward genetics revealed that CD36 is an accessory molecule for TLR2 in its response to diacylated lipopeptide and lipoteichoic acid (Hoebe et al. 2005). It is interesting to note that the mechanism by which CD36 functions as a TLR2 agonist is not elucidated yet. To date, it is not clear whether the TLR1/TLR2 complex also requires accessory molecules to be able to respond to triacylated lipopeptides.

4 BLPs and Human Innate Immune Proteins

BLPs, however, are amphipathic molecules comprising hydrophobic lipids bound to a hydrophilic part consisting of the polypeptide chain. Therefore, they are never found as a monomer in physiological solutions (Schromm et al. 2007). Instead they form either heterogeneous aggregates or micelle-like structures. These aggregations will make the molecules inaccessible to the immune system, in particular, to the ectodomain of TLR2, which does not have an apparent hydrophobic pocket. Interestingly, bacteria shed lipid complexes in vesicles that are very rich in BLPs, and these vesicles are likely to be the most relevant source of PAMPs in bacterial infections (Bhatnagar et al. 2007).

In the host, BLPs interact with carrier molecules to prevent their aggregation. Therefore, carrier proteins act as accessory molecules that regulate the presentation of BLP to TLR2. These carrier proteins are responsible for extracting BLPs from the supramolecular vesicles and present them to the surface of cells from the innate immune system. We (and others) have observed that BLP-mediated TLR2 activation requires serum (data not shown). Serum contains the major components of the humoral arm of the immune system as well as a plethora of proteins and lipoproteins that regulate the presence of lipids not only in serum but in the whole organism. Most likely these soluble proteins have the capacity of changing the fate of the BLPs within the host. To date, the humoral components reported to modulate TLR2 responses include LPS-binding protein (LBP) and soluble TLR2.

LBP is a protein with sequence similarity and conserved tertiary structure to other LPS-binding proteins such as bacterial/permeability-increasing protein (BPI). BPI and LBP function as components of the humoral arm of the immune system, and they mediate the host's response to LPS, although they are considered to have antagonistic functions. LBP alarms the host to the presence of LPS and can therefore be considered proinflammatory, while BPI renders LPS non-inflammatory (Weiss 2003). LBP has been demonstrated to bind to LPS monomers and present them to the TLR4 coreceptor CD14, acting upstream from TLR4 (Gioannini et al. 2002). It is noteworthy to mention that CD14 and LBP do not exhibit exclusive specificity for LPS, and many other bacterial products have been reported to interact with these proteins, including the BLPs.

Lower concentrations of LBP, which is an acute phase protein present in the bloodstream, enhances sensitivity of diacylated lipoproteins by catalyzing the transfer of lipoproteins to CD14, a glycosylphosphatidylinositol (GPI)-anchored protein expressed on the cell surface of phagocytes (Jiang et al. 2005; Schroder et al. 2004). CD14–lipoprotein complexes have been reported to facilitate the activation of TLR2 (Jiang et al. 2005; Sellati et al. 1998). However, surprisingly, mice succumb to the lethality of BLP-induced uncontrolled proinflammatory responses after blockade of LBP or CD14 by neutralizing antibodies (Heumann et al. 2003), suggesting that LBP and CD14 serve a scavenger function for BLPs in mice. Therefore, host responses to BLPs are not identical to that of LPS.

In addition, the interaction of lipoproteins with TLR2 has been proposed to be modulated by the existence of soluble forms of TLR2 in human plasma and breast milk (LeBouder et al. 2003), but the mechanism underlying this modulatory effect of soluble TLR2 is not fully defined. In the search for other carrier molecules in serum, we have developed affinity pull-downs of soluble-tagged BLPs. In these experiments, we have found that BLP interacts with many transport proteins in serum, including apolipoproteins and elements of the complement cascade. We are currently evaluating the relevance of such interactions.

5 Conclusions

The cells of the immune system are constantly exposed to potential pathogens and support large commensal bacteria, which are immunologically silent to the host response. The innate immune system, acting as the first line of defense, plays a pivotal role in maintaining such host–microbe homeostasis. Many of the molecules of the innate immunity recognize structures of the bacterial surface. Proteins crucial to the mediation of signals from the bacterial surface and the cells of the immune system are the humoral components of innate immune response and the TLRs.

How the various TLRs specifically bind and recognize their bacterial ligands is still not fully defined. For TLR4, after binding LPS, TLR4 forms a complex through interaction with its coreceptor, MD-2 (Kobayashi et al. 2006). TLR2 forms constitutive complexes with TLR1 and TLR6 to detect bacterial lipoproteins (Ozinsky et al. 2000). As for TLR9, binding of bacterial CpG DNA to preexisting dimmers of TLR9 is sufficient to induce activation (Latz et al. 2007). However, to date no studies on the formation of TLR oligomers have excluded the possibility of the existence of the combined activation of accessory soluble and cell-bound receptors that can result in complementary, synergistic, or antagonistic effects of the TLR signaling. Indeed, there has been growing evidence over the last few years that TLR responses are modulated by soluble endogenous molecules as well as integrins CD11c and/or CD18 (Kagan and Medzhitov 2006) and scavenger receptors such as CD36 (Hoebe et al. 2005). Endogenous molecules include S100A8 proteins (Vogl et al. 2007) and acute phase reactants such as LBP, soluble CD14, and pentraxins (Jeannin et al. 2005) among others. The S100A8 molecules, which are abundant secreted proteins of phagocytes, seem to mediate inflammatory responses only in combination with bacterial stimuli such as LPS. Whether this activity requires a direct interaction of LPS with these proteins or rather a synergistic effect of these molecules on the LPS–TLR4–MD2 complexes is a subject for further investigation. The long pentraxin PTX-3 is a multifunctional soluble pattern recognition receptor of the humoral arm of the immune system involved in tuning TLR2-mediated inflammation (Mantovani et al. 2007). PTX-3 recognizes selected pathogens and facilitates their phagocytosis and killing (Garlanda et al. 2002). Additionally, PTX-3 recognizes microbial moieties, among them the outer membrane protein A from *Klebsiella pneumoniae* (KpOmpA), a major component of the outer membrane of gram-negative bacteria highly conserved in the *Enterobacteriaceae* family (Jeannin et al. 2005). The interaction of PTX-3 with KpOmpA activates phagocytes in a TLR2-dependent way. However, the mechanism by which PTX-3 tunes the host response to KpOmpA has not been completely elucidated yet.

Strategies that target the TLR regulatory molecules might represent a new option for antiinflammatory therapies. Therefore further investigations should be performed to gain a better understanding of how these molecules modulate uncontrolled inflammatory processes induced by TLR-dependent activation.

References

Akira S, Uematsu S, Takeuchi O (2006) Pathogen recognition and innate immunity. Cell 124:783–801

Aliprantis AO, Yang RB, Mark MR, Suggett S, Devaux B, Radolf JD, Klimpel GR, Godowski P, Zychlinsky A (1999) Cell activation and apoptosis by bacterial lipoproteins through toll-like receptor-2. Science 285:736–739

Barrat FJ, Meeker T, Gregorio J, Chan JH, Uematsu S, Akira S, Chang B, Duramad O, Coffman RL (2005) Nucleic acids of mammalian origin can act as endogenous ligands for Toll-like receptors and may promote systemic lupus erythematosus. J Exp Med 202:1131–1139

Bhatnagar S, Shinagawa K, Castellino FJ, Schorey JS (2007) Exosomes released from macrophages infected with intracellular pathogens stimulate a proinflammatory response in vitro and in vivo. Blood 110:3234–3244

Brightbill HD, Libraty DH, Krutzik SR, Yang RB, Belisle JT, Bleharski JR, Maitland M, Norgard MV, Plevy SE, Smale ST, Brennan PJ, Bloom BR, Godowski PJ, Modlin RL (1999) Host defense mechanisms triggered by microbial lipoproteins through toll-like receptors. Science 285:732–736

Brown GD (2006) Dectin-1: a signalling non-TLR pattern-recognition receptor. Nat Rev Immunol 6:33–43

Buwitt-Beckmann U, Heine H, Wiesmuller KH, Jung G, Brock R, Ulmer AJ (2005) Lipopeptide structure determines TLR2 dependent cell activation level. Febs J 272:6354–6364

Buwitt-Beckmann U, Heine H, Wiesmuller KH, Jung G, Brock R, Akira S, Ulmer AJ (2006) TLR1- and TLR6-independent recognition of bacterial lipopeptides. J Biol Chem 281:9049–9057

Cohen J (2002) The immunopathogenesis of sepsis. Nature 420:885–891

Diebold SS, Kaisho T, Hemmi H, Akira S, Reis e Sousa C (2004) Innate antiviral responses by means of TLR7-mediated recognition of single-stranded RNA. Science 303:1529–1531

Foster SL, Hargreaves DC, Medzhitov R (2007) Gene-specific control of inflammation by TLR-induced chromatin modifications. Nature 447:972–978

Garlanda C, Hirsch E, Bozza S, Salustri A, De Acetis M, Nota R, Maccagno A, Riva F, Bottazzi B, Peri G, Doni A, Vago L, Botto M, De Santis R, Carminati P, Siracusa G, Altruda F, Vecchi A, Romani L, Mantovani A (2002) Non-redundant role of the long pentraxin PTX3 in anti-fungal innate immune response. Nature 420:182–186

Gerold G, Zychlinsky A, de Diego JL (2007) What is the role of Toll-like receptors in bacterial infections? Semin Immunol 19:41–47

Gioannini TL, Zhang D, Teghanemt A, Weiss JP (2002) An essential role for albumin in the interaction of endotoxin with lipopolysaccharide-binding protein and sCD14 and resultant cell activation. J Biol Chem 277:47818–47825

Guo L, Lim KB, Gunn JS, Bainbridge B, Darveau RP, Hackett M, Miller SI (1997) Regulation of lipid A modifications by Salmonella typhimurium virulence genes phoP-phoQ. Science 276:250–253

Heumann D, Lauener R, Ryffel B (2003) The dual role of LBP and CD14 in response to Gram-negative bacteria or Gram-negative compounds. J Endotoxin Res 9:381–384

Hoebe K, Georgel P, Rutschmann S, Du X, Mudd S, Crozat K, Sovath S, Shamel L, Hartung T, Zahringer U, Beutler B (2005) CD36 is a sensor of diacylglycerides. Nature 433:523–527

Jeannin P, Bottazzi B, Sironi M, Doni A, Rusnati M, Presta M, Maina V, Magistrelli G, Haeuw JF, Hoeffel G, Thieblemont N, Corvaia N, Garlanda C, Delneste Y, Mantovani A (2005) Complexity and complementarity of outer membrane protein A recognition by cellular and humoral innate immunity receptors. Immunity 22:551–560

Jiang Z, Georgel P, Du X, Shamel L, Sovath S, Mudd S, Huber M, Kalis C, Keck S, Galanos C, Freudenberg M, Beutler B (2005) CD14 is required for MyD88-independent LPS signaling. Nat Immunol 6:565–570

Kagan JC, Medzhitov R (2006) Phosphoinositide-mediated adaptor recruitment controls Toll-like receptor signaling. Cell 125:943–955

Karin M, Lawrence T, Nizet V (2006) Innate immunity gone awry: linking microbial infections to chronic inflammation and cancer. Cell 124:823–835

Kobayashi M, Saitoh S, Tanimura N, Takahashi K, Kawasaki K, Nishijima M, Fujimoto Y, Fukase K, Akashi-Takamura S, Miyake K (2006) Regulatory roles for MD-2 and TLR4 in ligand-induced receptor clustering. J Immunol 176:6211–6218

Krutzik SR, Ochoa MT, Sieling PA, Uematsu S, Ng YW, Legaspi A, Liu PT, Cole ST, Godowski PJ, Maeda Y, Sarno EN, Norgard MV, Brennan PJ, Akira S, Rea TH, Modlin RL (2003) Activation and regulation of Toll-like receptors 2 and 1 in human leprosy. Nat Med 9:525–532

Latz E, Verma A, Visintin A, Gong M, Sirois CM, Klein DC, Monks BG, McKnight CJ, Lamphier MS, Duprex WP, Espevik T, Golenbock DT (2007) Ligand-induced conformational changes allosterically activate Toll-like receptor 9. Nat Immunol 8:772–779

Leadbetter EA, Rifkin IR, Hohlbaum AM, Beaudette BC, Shlomchik MJ, Marshak-Rothstein A (2002) Chromatin-IgG complexes activate B cells by dual engagement of IgM and Toll-like receptors. Nature 416:603–607

LeBouder E, Rey-Nores JE, Rushmere NK, Grigorov M, Lawn SD, Affolter M, Griffin GE, Ferrara P, Schiffrin EJ, Morgan BP, Labeta MO (2003) Soluble forms of Toll-like receptor (TLR)2 capable of modulating TLR2 signaling are present in human plasma and breast milk. J Immunol 171:6680–6689

Lee MS, Kim YJ (2007) Signaling pathways downstream of pattern-recognition receptors and their cross talk. Annu Rev Biochem 76:447–480

Liew FY, Xu D, Brint EK, O'Neill LA (2005) Negative regulation of toll-like receptor-mediated immune responses. Nat Rev Immunol 5:446–458

Lorenz E, Mira JP, Cornish KL, Arbour NC, Schwartz DA (2000) A novel polymorphism in the toll-like receptor 2 gene and its potential association with staphylococcal infection. Infect Immun 68:6398–6401

Lotze MT, Tracey KJ (2005) High-mobility group box 1 protein (HMGB1): nuclear weapon in the immune arsenal. Nat Rev Immunol 5:331–342

Mantovani A, Garlanda C, Doni A, Bottazzi B (2007) Pentraxins in innate immunity: from C-reactive protein to the long pentraxin PTX3. J Clin Immunol Sep 9 [Epub ahead of print]

Medzhitov R, Janeway C Jr (2000) Innate immune recognition: mechanisms and pathways. Immunol Rev 173:89–97

Medzhitov R, Janeway CA Jr (1998) Innate immune recognition and control of adaptive immune responses. Semin Immunol 10:351–353

Ogus AC, Yoldas B, Ozdemir T, Uguz A, Olcen S, Keser I, Coskun M, Cilli A, Yegin O (2004) The Arg753GLn polymorphism of the human toll-like receptor 2 gene in tuberculosis disease. Eur Respir J 23:219–223

Ozinsky A, Underhill DM, Fontenot JD, Hajjar AM, Smith KD, Wilson CB, Schroeder L, Aderem A (2000) The repertoire for pattern recognition of pathogens by the innate immune system is defined by cooperation between toll-like receptors. Proc Natl Acad Sci USA 97:13766–13771

Randow F, Seed B (2001) Endoplasmic reticulum chaperone gp96 is required for innate immunity but not cell viability. Nat Cell Biol 3:891–896

Roach JC, Glusman G, Rowen L, Kaur A, Purcell MK, Smith KD, Hood LE, Aderem A (2005) The evolution of vertebrate Toll-like receptors. Proc Natl Acad Sci USA 102:9577–9582

Schroder NW, Heine H, Alexander C, Manukyan M, Eckert J, Hamann L, Gobel UB, Schumann RR (2004) Lipopolysaccharide binding protein binds to triacylated and diacylated lipopeptides and mediates innate immune responses. J Immunol 173:2683–2691

Schromm AB, Howe J, Ulmer AJ, Wiesmuller KH, Seyberth T, Jung G, Rossle M, Koch MH, Gutsmann T, Brandenburg K (2007) Physicochemical and biological analysis of synthetic bacterial lipopeptides: validity of the concept of endotoxic conformation. J Biol Chem 282:11030–11037

Sellati TJ, Bouis DA, Kitchens RL, Darveau RP, Pugin J, Ulevitch RJ, Gangloff SC, Goyert SM, Norgard MV, Radolf JD (1998) Treponema pallidum and Borrelia burgdorferi lipoproteins and synthetic lipopeptides activate monocytic cells via a CD14-dependent pathway distinct from that used by lipopolysaccharide. J Immunol 160:5455–5464

Srivastava P (2002) Roles of heat-shock proteins in innate and adaptive immunity. Nat Rev Immunol 2:185–194

Takeuchi O, Kaufmann A, Grote K, Kawai T, Hoshino K, Morr M, Muhlradt PF, Akira S (2000) Cutting edge: preferentially the R-stereoisomer of the mycoplasmal lipopeptide macrophage-activating lipopeptide-2 activates immune cells through a toll-like receptor 2- and MyD88-dependent signaling pathway. J Immunol 164:554–557

Takeuchi O, Kawai T, Muhlradt PF, Morr M, Radolf JD, Zychlinsky A, Takeda K, Akira S (2001) Discrimination of bacterial lipoproteins by Toll-like receptor 6. Int Immunol 13:933–940

Takeuchi O, Sato S, Horiuchi T, Hoshino K, Takeda K, Dong Z, Modlin RL, Akira S (2002) Cutting edge: role of Toll-like receptor 1 in mediating immune response to microbial lipoproteins. J Immunol 169:10–14

Trinchieri G, Sher A (2007) Cooperation of Toll-like receptor signals in innate immune defence. Nat Rev Immunol 7:179–190

Vogl T, Tenbrock K, Ludwig S, Leukert N, Ehrhardt C, van Zoelen MA, Nacken W, Foell D, van der Poll T, Sorg C, Roth J (2007) Mrp8 and Mrp14 are endogenous activators of Toll-like receptor 4, promoting lethal, endotoxin-induced shock. Nat Med 13:1042–1049

Weiss J (2003) Bactericidal/permeability-increasing protein (BPI) and lipopolysaccharide-binding protein (LBP): structure, function and regulation in host defence against Gram-negative bacteria. Biochem Soc Trans 31:785–790

West AP, Koblansky AA, Ghosh S (2006) Recognition and signaling by Toll-like receptors. Annu Rev Cell Dev Biol 22:409–437

Yang Y, Liu B, Dai J, Srivastava PK, Zammit DJ, Lefrancois L, Li Z (2007) Heat shock protein gp96 is a master chaperone for toll-like receptors and is important in the innate function of macrophages. Immunity 26:215–226

Ernst Schering Foundation Symposium Proceedings, Vol. 3, pp. 97–137
DOI 10.1007/2789_2007_062

Published Online: 18 December 2007

Migration, Cell–Cell Interaction and Adhesion in the Immune System

M. Gunzer(✉)

Institute of Immunology, Otto-von-Gericke University Magdeburg,
Leipziger Straße 44, D-39120 Magdeburg, Germany
email: *Matthias.gunzer@med.ovgu.de*

Abstract. Migration is an essential function of immune cells. It is necessary to lead immune cell precursors from their site of generation to the places of maturation or function. Cells of the adaptive immune system also need to interact physically with each other or with specialized antigen presenting cells in lymphatic tissues in order to become activated. Thereby a complex series of controlled migration events, adhesive interactions and signalling responses is induced. Finally cells must be able to leave the activating tissues and re-enter the bloodstream from which they extravasate into inflamed tissue sites. Cells of the innate immune system can function directly without the need for previous activation. However, these cells have to adapt their function to a panoply of pathogens and environmental niches which can be invaded. The current review highlights the central aspects of cellular dynamics underlying adaptive and innate cellular immunity. Thereby a focus will be put on recent results obtained by microscopic observation of live cells in vitro or by intravital 2-photon microscopy in live animals.

1 Introduction

The ability of immune cells to migrate autonomously is an essential prerequisite for a functioning immune system. Thereby the movement of precursor cells from their origins of generation into secondary developmental sites, such as the thymus for T cells or lymph nodes for B cells, is one important arm of immunological function that critically depends on autonomous cell migration. Immune cells have to migrate into lymphatic tissues and manoeuvre within in order to achieve activation. In addition, a highly organized and controlled series of physical interactions between different types of immune cells is required for successful activation. These have to be established, maintained for a defined period of time by means of cell adhesion and communication, and ultimately resolved to produce freely mobile individual cells again. Finally, many effector cells leave the site of activation and enter areas of direct action such as inflamed wounds. Again, this requires controlled migration as well as adhesion to vessel walls and endothelial cells for transmigration into the inflicted tissues. This review attempts to highlight the major sites of immune cell migration and the approaches for their investigation with a focus on mature T and B lymphocytes.

2 Migration of Haematopoietic Stem Cells

Already during the development of the immune system in the embryo massive movements of haematopoietic precursor cells from the yolk sac to the foetal liver and from there into the areas of blood formation in the bone marrow cavities are required (Godin and Cumano 2005). The movement of stem cells within the bone marrow itself is still poorly understood. First results suggest that haematopoietic stem cells are located in close proximity to the bone matrix which is rich in calcium (Adams et al. 2006; Suzuki et al. 2006). It is assumed that during events which require the release of mature cells from the marrow, such as peripheral infection or heavy bleeding, haematopoietic stem cells move from the peripheral zone of the bone into the more vascularized centre and proliferate in this region. Differentiated daughter cells generated during this reaction can cross the vessel wall and enter the free blood stream (Wilson and Trumpp 2006). By this mechanism all leukocytes and erythrocytes are being released from the bone marrow.

3 Migration Processes During B Cell Development

3.1 Initial Activation of B Cells

While granulocytes and monocytes are mature and functional directly after their release from the bone marrow, both B and T lymphocytes must perform additional maturation steps in the periphery. To this end newly released B cells immigrate into peripheral lymphatic organs where they wait until they can bind an antigen (typically proteins, but also any other defined chemical entity such as sugars or small molecules) in its naturally occurring three-dimensional conformation via the B cell receptor (BCR). The BCR is a membrane-bound version of the antibody molecule which is being produced by this particular B cell. Binding of antigen to the BCR pre-activates the B cell, leads to its proliferation and allows the release of multiple copies of the BCR in a soluble version.

B cells secreting large amounts of antibodies are called plasma cells. Since the recognition site of the BCR is generated by chance recombination of a large collection of genetic building blocks (Matthias and Rolink 2005) only very few B cell clones are able to specifically react to a given antigen. The antigen-dependent activation of B cells ensures that only the right B cells proliferate and, therefore, only cognate antibody molecules are produced.

For full activation and change of the class of antibodies from the initial dekavalent IgM to the divalent IgG/IgE or the tetravalent IgA molecules, B cells also need the physical contact of activated T helper cells. Via its T cell receptor (TCR) the T helper cell recognizes peptide fragments of the antigens, which the B cell has taken up by means of its BCR and is presenting in the peptide-binding grove of MHC II molecules on its surface. This recognition leads to up-regulation of the CD40-ligand molecule on T helper cells which in turn induces antibody class switch via stimulation of the CD40 receptor on the B cell surface (Okada and Cyster 2006; Banchereau et al. 1994). These interactions between B cells and helper T cells occur in the areas of a B cell follicle where it is close to the T cell zone (Garside et al. 1998; Reif et al. 2002; Manser 2004). To reach the follicular border and come into close proximity with T cells, B cells undergo a change in chemokine receptor expression after BCR triggering. As resting B cells they express the receptor CXCR5, which recognizes the follicular chemokine CXCL13 (Förster et al. 1996) and thus keeps the cells in the central area of the follicle. After BCR-mediated activation B cells undergo a switch in chemokine receptor expression to CCR7. This receptor allows B cells to detect the T zone chemokines CCL19 and CCL21 and to migrate towards the source of these chemokines (Förster et al. 1999; Reif et al. 2002; Okada et al. 2005). The switch in chemokine receptor expression is rapidly induced after BCR triggering (Okada et al. 2005).

3.2 The Germinal Centre Reaction

The antibody molecules produced in the early phase of an adaptive immune response show relatively low affinity towards the inducing antigen. In the course of an infection and at the time of a secondary exposure to the same antigen, antibodies with much higher affinity are being produced. This process has been termed affinity maturation (Aydar et al. 2005). The basis of affinity maturation is the germinal centre reaction (Moser et al. 2006). After primary activation via antigen and T helper cells, fully activated B cells migrate back into the centre of a B cell follicle and form a germinal centre (GC) (MacLennan 1994). The GC is structured into a dark zone, where highly proliferative centroblasts are engaged in producing variants of the original BCR by means of random site directed V-gene mutation (MacLennan 1994). Centroblasts then move into the light zone, where they are called centrocytes. The light zone is rich in a specific cell type, the follicular dendritic cell (FDC) (Tarlinton 1998). On their cell surface FDC present immunecomplexes (iccosomes) consisting of specific antibodies and bound antigen (Kosco-Vilbois 2003). Centrocytes test the affinity of their newly developed antibody molecules via physical contact to the iccosomes of FDC. Those B cells that have generated antibodies with higher affinity receive a survival signal, while cells that have generated an antibody with weaker affinity undergo apoptosis and are removed by tingible body macrophages (MacLennan 1994). As a result of this process the affinity of the secreted antibodies is steadily rising. GC-derived plasma cells can either stay in the medulla of lymph nodes or enter specialized niches in the bone marrow as long-lived plasma cells. They can also migrate to special areas of the body where an infection is taking place to release the antibodies locally (Moser et al. 2006).

Until 2 years ago the dynamics of the migratory processes underlying the onset of a germinal centre were unknown. Only histological studies were able to describe the sequential steps of the process (Garside et al. 1998; Reif et al. 2002; Cyster 1999; Moser et al. 2006; MacLennan 1994; de Vinuesa et al. 2000). Then an initial study showed the dynamics of B cells directly after BCR stimulation by time-lapse 2-photon microscopy (2PM) in intact lymph nodes (Okada et al. 2005). In the past several months two studies have independently investigated the migratory behaviour and contact formation of centrocytes, centroblasts, FDC and follicular T helper cells by 2PM either in explanted lymph nodes or in vivo in mice (Allen et al. 2007; Schwickert et al. 2007). Both studies demonstrated the enormous migratory activity of centrocytes. Interestingly, in the presence of specific antigen, the migration of centrocytes, although equally fast, was much more confined as compared to the migration of unrelated follicular B cells (Schwickert et al. 2007). Contacts to cognate FDC were only brief (in the range of a few minutes) but significantly longer than with non-specific B cells (Schwickert et al. 2007). Both studies also detected migration of cells from the dark to the light zone but also in the reverse direction, suggesting a repeated hypermutation-FDC probing behaviour in the GC reaction (Schwickert et al. 2007; Allen et al. 2007), which was previously only speculated upon.

The entire GC reaction can last for several weeks, a period which is far too long for permanent imaging. Thus, in order to fully understand the cellular migration events during the entire GC reaction, imaging snapshots like the ones described above need to be analysed by mathematical modelling approaches. Only then it will be possible to extrapolate the highly complex and delicately balanced long-term migratory and cell-contact processes associated with the entire GC reaction (Meyer-Hermann and Maini 2005).

4 Migration Processes During T Cell Development

4.1 Thymic Maturation and Antigen Presentation

An equally complex pattern of migration and cell–cell interaction forms the basis of T cell development. Immature T cell progenitors leave the bone marrow and immigrate into the thymus (Boehm and Bleul 2006). Within the thymus progenitor cells transform into thymocytes which migrate from the peripheral parts of the cortex into the medulla. Thereby they engage tightly with thymic epithelial cells and thymic dendritic cells (DC). After having crossed the entire thymus, thymocytes have completed their development into mature naïve T cells and extravasate via the blood vessel system (Reichardt and Gunzer 2006; Bousso and Robey 2004).

This process is very important, as it not only matures thymocytes into real T cells but also makes sure that those T cells that then leave the thymus are not autoaggressive. The TCR of thymocytes is generated by a mechanism very similar to the one established for the generation of the BCR. Multiple genetic segments can be freely recombined by developing thymocytes. Thus, in principle a TCR with the capability to react to self structures can be formed. The principal difference between BCR and TCR is the fact that the latter can recognize only linear peptides derived from the primary sequence of a protein. These peptides have to be physically associated with the MHC complexes of antigen presenting cells or normal body cells (Babbitt et al. 1985; Bjorkman et al. 1987a,b). The binding of foreign peptides to self MHC molecules has been termed 'altered self', where the MHC is 'self' and the alteration is induced by the association of MHC with foreign peptides (Zinkernagel and Doherty 1974).

Peptide–MHC complexes can be formed either by specific antigen presenting cells (APC) after uptake of foreign proteins or by production of peptide fragments from endogenously produced proteins by all cells of the body. The process of peptide–MHC generation is very complex (Benham et al. 1995; Trombetta and Mellman 2005). Peptides are generated via proteolytic breakdown of proteins. In a highly regulated process these peptides bind to unloaded MHC molecules in specific loading compartments within the cytoplasm. Loaded MHC molecules then move to the cell surface of the presenting cell. Endogenous peptides are normally associated with MHCI molecules and peptides derived from ingested foreign proteins are associated with MHC II complexes. Peptide–MHC I complexes are recognized by $CD8^+$ cytotoxic T cells and peptide–MHC II complexes by $CD4^+$ helper T cells (Chow and Mellman 2005). There is also a process termed cross-presentation. Here, peptides derived from exogenous proteins are presented in MHC I (Norbury et al. 2004; Wolkers et al. 2004; Ackerman and Cresswell 2004). All nucleated body cells carry MHCI molecules; however, MHC II molecules are expressed only by APC.

Antigen presentation in the thymus is unique since APC in this organ present self peptides exclusively, both in MHC I and MHC II. Thymic selection has two main goals. One is the elimination of all thymocytes which are not able to recognize self MHC molecules at all, since the configuration of their TCR is unable to bind to self MHC. This is achieved by providing a survival signal only to thymocytes that bind self MHC with a certain affinity (positive selection). In the second step all those thymocytes are eliminated whose binding to self peptide–MHC complexes is too efficient. Such cells bear the risk of being autoaggressive (negative selection; von Boehmer et al. 2003). Within the thymus double-positive CD4/CD8 thymocytes also decide whether they become CD4 helper cells or CD8 cytotoxic T cells (Germain 2002). In summary, the thymus generates a T cell pool with a widespread potential reactivity against foreign structures in the presence of an almost complete self tolerance.

4.2 Cellular Dynamics of Thymic Selection

Microscopic analyses in thymic reaggregate cultures showed highly dynamic motility of thymocytes during positive selection (Bousso et al. 2002). Fifty per cent of contacts with stromal cells (and possibly also local DC) were short-lived and dynamic, 50% were more stable. However, until today no functional difference has been associated with those different interaction kinetics. Experiments in intact thymic slices showed that intracellular calcium plays a major role in controlling the migration of thymocytes. In these experiments increases in calcium were associated with breaks in migration while shortly before onset of motile phases thymocytes down-regulated their intracellular calcium levels (Bhakta et al. 2005). The beginning of positive selection, that is the first functional contacts with local APC, seems to transform a previously random migration behaviour of thymocytes to a highly oriented migration into the medulla of the thymus (Witt et al. 2005b). It is speculated that a change in chemokine receptor expression similar to that described above for BCR-stimulated B cells also underlies the change in orientation of thymocytes (Reif et al. 2002; Witt et al. 2005a; Misslitz et al. 2004).

5 Migration Processes During T Cell Activation

5.1 Activation of Naïve T Cells in the Lymph Node

Thymocytes which have survived all processes of selection in the thymus enter the blood stream as mature naïve T cells. They are now searching for APC which present peptide–MHC complexes that fit to the specificity of their TCR. This poses an enormous logistical problem. Presentation of peptide–MHC complexes requires the physical interaction of APC and T cells. APC, typically DC, are located in every tissue of the body, specifically at the surfaces interacting with the environment such as skin or mucosa. Here they take up foreign antigens and present them in their MHC molecules. Compared to the size of a single leukocyte the body cavity is a huge universe. Thereby, T cells as well as APC are migrating independently in different areas of the body, have no information on the location of the other cell and are unable to migrate specifically towards each other. This requires a mechanism which focuses random motility of all cells in a way that finally allows specific T cell activation by physical T cell–APC interaction.

This is made possible by a system that, based on its principles of function can be ideally described with a marketplace analogy (Fig. 1; Reichardt and Gunzer 2006). A marketplace is a central, well-described place where a large number of merchants offer a wide variety of goods. Customers visiting the marketplace have certain requirements which they hope to meet by visiting the marketplace. Since they are not able to tell which merchant will sell the relevant goods, they must visit as many merchants as possible, ideally all. They will finally make a deal with the merchant(s) who serve(s) their needs most efficiently. Transferred to T cell activation the marketplace is provided by a lymphatic organ, DC are the merchants who sell their 'goods', peptide–MHC complexes. T cells are customers.

In reality T cells leave the blood vessels in the areas of high endothelial venules within lymph nodes (von Andrian and Mempel 2003). After crossing the endothelial layer T cells move to deeper regions of the T cell zones (von Andrian and Mempel 2003; Warnock et al. 1998). Along their paths T cells engage in physical contacts with surrounding DC to test the antigens that are presented (Germain and Jenkins 2004). In contrast, DC migrate from peripheral zones of the body, for example the skin, where they take up, process and present foreign antigen in their MHC complexes (Banchereau et al. 2000; Reis e Sousa 2006; Gunzer et al. 2001), into T zones of lymph nodes. Thus, the lymph node is a melting pot for migratory cells from multiple body sites. Just by this movement of cells into the lymph node a concentration by a factor of 10,000–30,000 is obtained. Thus, the search space for cells is condensed from the entire body volume into the volume of a pinhead. But also a pinhead is a large volume for T cells or DC also. A normal mouse lymph node contains approximately 1×10^6 cells. Thus within

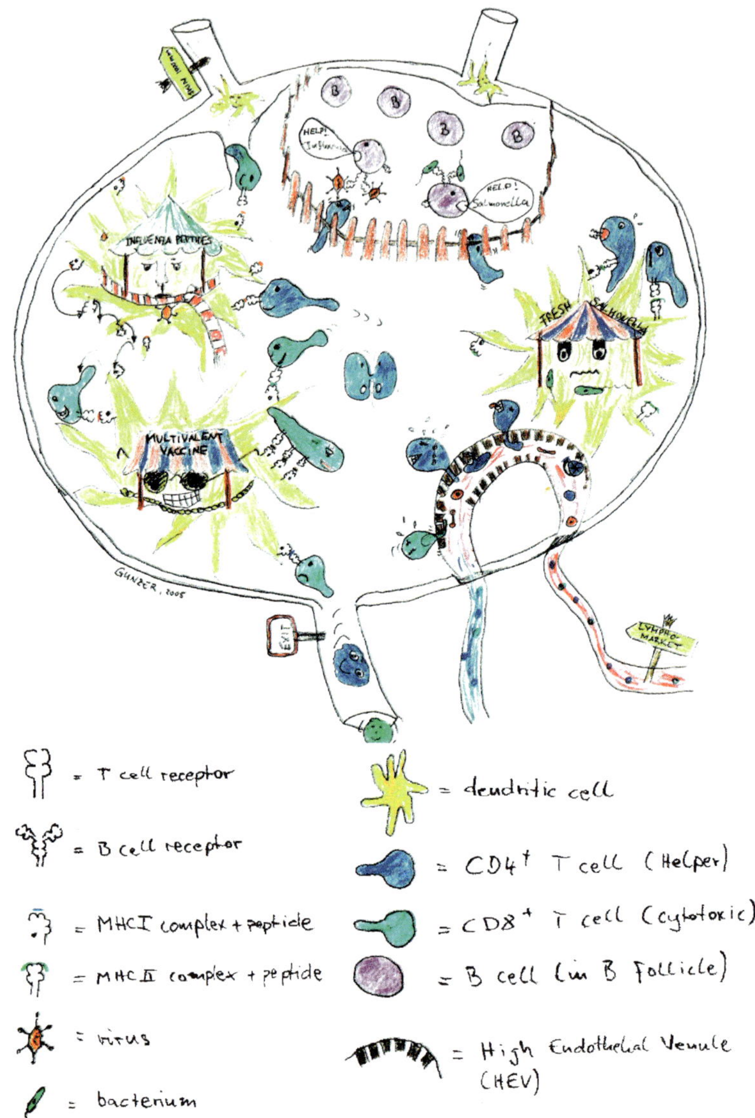

B
HELP! Influenza
HELP! Salmonella
INFLUENZA PEPTIDES
FRESH SALMONELLA
MULTIVALENT VACCINE
EXIT
LYMPHO-MARKET
GUNZER, 2005
= T cell receptor
= B cell receptor
= MHCI complex + peptide
= MHCII complex + peptide
= virus
= bacterium
= dendritic cell
= CD4+ T cell (Helper)
= CD8+ T cell (cytotoxic)
= B cell (in B Follicle)
= High Endothelial Venule (HEV)

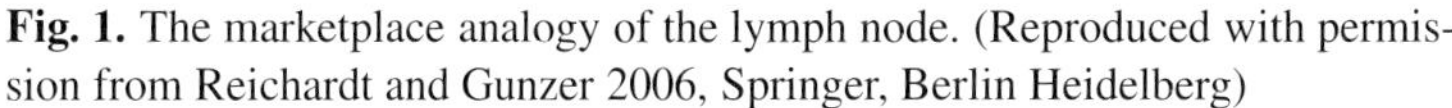

Fig. 1. The marketplace analogy of the lymph node. (Reproduced with permission from Reichardt and Gunzer 2006, Springer, Berlin Heidelberg)

the lymph node also, the migration of cells needs to be controlled in order to enhance the likelihood of their meeting (Reichardt and Gunzer 2006).

5.2 Dynamics of T Cell Activation in Lymph Nodes

T cell activation has traditionally been a field of major interest in immunological research. Nevertheless, the observation of individual cells and their migration during activation has received little attention for a long time. A study from the group of Alan Aderem showed for the first time the dynamics of T cells during activation in a two-dimensional system (i.e., a cell culture dish; Underhill et al. 1999). Here, a T cell clone was observed during its activation by a macrophage line as APC (Underhill et al. 1999). However, one major characteristic of true lymphatic tissue is its inherent three-dimensionality and primary activation of naïve T cells was a central question as this is induced by mature DC rather than macrophages.

The optimal method to observe cell migration under realistic conditions is intravital microscopy (von Andrian and M'Rini 1998). This, however, is technically challenging and was not established in many laboratories in the mid 1990s. A much simpler method to of obtaining a three-dimensional environment with many characteristics of true extracellular matrix is the use of hydrated gels of type I collagen as an environment for cell migration. Such gels are transparent and migrating cells can be observed easily by conventional wide-field microscopy (Friedl et al. 1993; Reichardt et al. 2007b). Using three-dimensional collagen gels allowed the study of T cell migration and interaction with APC by us (Friedl et al. 1995; Gunzer et al. 1997) and by other groups (Huang et al. 2005). Especially, it enabled us to observe the dynamics of naïve T cells during antigen specific activation by mature DC in a three-dimensional environment (Gunzer et al. 2000b). The major advancement was the use of naïve T cells from the TCR-transgenic mouse line DO11.10, that carries a fully recombined TCR against a peptide of chicken ovalbumin (Murphy et al. 1990). To activate these cells we used mature DC from the bone marrow of mice (Gunzer et al. 2000b). It became obvious that the transition from a two-dimensional into a three-dimensional system induced drastic changes in cellular motility and, specifically, cell–cell interactions (Gunzer et al. 2000b). We could not observe large cellular aggregates of DC and T cells any more, which are very common in two-dimensional systems (Gunzer et al. 2000b). On the single cell level we could observe highly dynamic and short-lived interactions of T cells with DC despite potent T cell activation (Gunzer et al. 2000b). This study also introduced the parameter of T cell–DC contact duration which was later used in many follow-up studies.

The first analyses of T cell migration in intact lymphatic tissues were published in 2002/2003. These showed that leukocytes migrate effectively in lymphatic tissues (Miller et al. 2002, 2004a; Bousso and Robey 2003; Stoll et al. 2002). T cells migrated faster than B cells or DC (Miller et al. 2002; Bousso and Robey 2003). Interestingly, the values measured in intact lymphatic tissue closely resembled those for the same types of cells obtained in three-dimensional collagen matrices (Friedl et al. 1993, 1994; Dorner et al. 1997; Entschladen et al. 1997; Friedl et al. 1998; Gunzer et al. 2000a, 2004).

Meanwhile T cell migration in lymphatic organs is an intensively studied topic in experimental immunology, especially since the introduction of 2PM in many laboratories. 2PM is able to visualize fluorescently labelled cells deep within lymphatic tissue, which is a prerequisite for the analysis of cell migration in situ (Cahalan et al. 2002). Published studies demonstrated the difference in the basal motility of DC in lymph nodes (Lindquist et al. 2004) or showed direct stimulatory contacts between immigrating B cells and antigen loaded DC (Qi et al. 2006) or the induction of chemotaxis in CD8 cells towards DC, which had previously been contacted by an antigen specific helper T cell (Castellino et al. 2006). In addition, a possible role of regulatory T cells in the suppression of cytotoxic T cells was demonstrated by means of intravital microscopy (Mempel et al. 2006). Meanwhile the number of new studies using intravital 2PM is too large to be covered here in its entirety.

5.3 Analysis of T Cell–APC Interactions: The Importance of Contact Times

The majority of previously published imaging studies in lymphatic tissue, however, were concerned with the analysis of contact times between T cells and APC, which were mostly DC. This is true for studies in explanted lymph nodes (Bousso et al. 2002; Miller et al. 2002, 2004a, 2004b; Bousso and Robey 2003; Stoll et al. 2002; Okada et al. 2005; Hugues et al. 2004; Tang et al. 2006) as well as for the first set of studies on fully vascularized lymph nodes in living animals (Mempel et al. 2004; Tadokoro et al. 2006). As mentioned above, the importance of T cell–APC contact times was first recognized in a paper from our group investigating the dynamics of T cells contacting DC in artificial three-dimensional matrix systems (Gunzer et al. 2000b). Based on kinetic analyses of T cell activation it had previously been shown that T cells must be in contact with DC for at least 6, or better 20 h to be fully activated (Iezzi et al. 1998). Although cells were not directly observed during the process it was assumed that the contact of T cells with APC had to be permanently maintained for the measured duration (Lanzavecchia et al. 1999). Thus, our demonstration that naïve T cells need to be able to contact DC for a certain period, and that such contacts were not permanent but rather transient and dynamic and the cells were permanently migrating over each other (Gunzer et al. 2000b) led to intensive discussions in the field (Lanzavecchia and Sallusto 2001a, 2001b; Dustin and Chan 2000; Dustin and de Fougerolles 2001; Dustin et al. 2001).

5.4 Models of Cellular Dynamics During Antigen Presentation

The observation of dynamic cell migration during antigen presentation prompted us to develop the 'serial encounter model of T cell activation' (Friedl and Gunzer 2001). The model states that T cells in the lymph node are generally dynamic, independent of whether they 'see' their antigen or not. Only the fact that T cells receive a specific, activating and proliferation-inducing signal during a specific contact distinguishes specific from non-specific T–DC encounters. In addition, a single specific contact would be too short to transport all necessary information for full T cell activation. Thus, naïve T cells are forced to engage in a series of productive contacts and sum up the signal coming from these, thus the name serial encounter model. As a result the density of antigen loaded APC in the migratory reach of T cells would be a decisive factor for the induction of a reaction. Responses against antigens which reach only a low density and, consequently, induce only a low number of loaded DC (e.g., at the end of an infection), would be rendered highly unlikely (Friedl and Gunzer 2001). This hypothesis made testable predictions which could only be answered by direct visual inspection of migrating cells within intact lymph nodes (Dustin et al. 2001).

Today there is little discussion about the existence of both transient and stable contacts between antigen specific naïve T cells and cognate DC. The sequence of events in the lymph node currently agreed upon is as follows: in the first 8 h of a T cell response towards a novel antigen, T cells are very dynamic and migrate efficiently within the lymph node parenchyma despite frequent contacts to cognate DC. Contacts then gradually get longer and at some time between 12 and 20 h of the response the majority of T–DC contacts last >60 min (Bousso and Robey 2003; Miller et al. 2004b; Mempel et al. 2004). Interestingly, this stable second phase could not be reconstructed in artificial three-dimensional collagen matrices. Thus, factors which are only present in intact lymph nodes seem to be responsible for this effect. A third phase is characterized by higher motility in T cells as well as short contacts with DC (Mempel et al. 2004). Maximal T cell proliferation begins at the end of phase 3, but maximal production of effector cytokines [interleukin (IL)-2 and interferon (IFN)]-γ] correlates with the end of phase 2 (Mempel et al. 2004). Of note, this simple scheme has recently been challenged by data showing that the inability of T cells to form stable contacts with cognate DC late in a priming reaction is due to the high density of adoptively transferred T cells in these studies rather than a principal defect of the T cells (Garcia et al. 2007). At 10-fold lower numbers of T cells, which seems to be more physiological, T cells in late phases also undergo a large number of stable contacts with DC. Thus, it seems necessary to develop novel tools to allow imaging of a situation as close as possible to the natural one to get an idea about the real dynamics of T cell activation in vivo.

5.5 The Plasticity of T Cell Effector Functions: Role of the APC

Despite the frequent observation of different contact kinetics in vitro and in vivo it is currently an open question whether these different kinetics influence the result of T cell activation, or whether naïve T cells always follow the same pattern of molecular events during activation, irrespective of the contact kinetics to APC. The Th1/Th2 paradigm of T cell activation (Mosmann et al. 1986) or the recent discovery of the Th17-type of T cells (Harrington et al. 2005) suggests that the activation of T cells allows a large spectrum of variations, at least on a functional level. Thus the question arises whether interactions between T cells and different APC also induce different T cell effector phenotypes. We have approached this problem by comparing directly the kinetics of T cells interacting with DC or B cells as APC. Given to the number of cells in the body B cells are by far the most abundant MHC II-positive APC in the body. Whether they take part in the primary activation of naïve T cells is still a matter of intense debate (Matzinger 1994; Epstein et al. 1995; Lassila et al. 1988). In fact, before the discovery of DC macrophages and B cells were considered the primary APC for T cells (Lipsky and Rosenthal 1975; Steinman 1991). This changed with the demonstration by Steinman that highly pure DC are the cell population in spleen which is able to induce maximal T cell proliferation in vitro (Steinman et al. 1983; Steinman and Witmer 1978). Although DC are considered to be the most efficient APC today (Banchereau and Palucka 2005) B cells are still associated with antigen presentation in many studies, either in a regulatory manner (Knoechel et al. 2005; Mizoguchi and Bhan 2006; Fuchs and Matzinger 1992; Fillatreau et al. 2002) or as real T cell stimulators (Crawford et al. 2006; Rodriguez-Pinto and Moreno 2005).

We had shown previously that naïve B cells can activate naïve T cells in an antigen specific manner. However, on a per cell basis the efficiency of B cells for T cell activation was 100-fold lower than that of DC (Gunzer et al. 2004). To our surprise, an investigation of the physical interactions between peptide-loaded naïve B cells and antigen-specific naïve T cells showed that the cells formed long-lived monogamous contacts. Thereby very characteristic patterns of T–B pairs were formed. The T

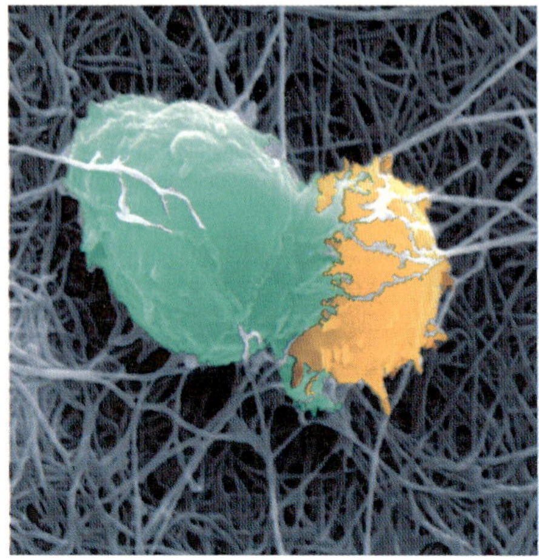

Fig. 2. A naïve DO11.10 T cell (*green*) interacts with a naïve B cell (*orange*) within a 3-D collagen matrix. Scanning electron micrograph of a fixed cell pair within a three-dimensional collagen matrix. The naïve B cell presents specific antigen towards the T cell. Note the formation of an intensive contact between the cells as well as the elongated (migratory) morphology of the T cell. Such cell pairs are highly motile in three-dimensional collagen matrices in vitro as well as in lymph nodes in vivo. Modified from research originally published in *Blood*. (Gunzer et al. 2004, *Blood*, © American Society of Hematology)

cell was always elongated and the B cell round (Fig. 2). The morphology of the T cell suggested that such cell pairs could also be motile. Indeed we could demonstrate that T–B pairs were migrating very efficiently both in artificial three-dimensional collagen matrices as well as in lymph nodes of live mice (Gunzer et al. 2004). This was later confirmed by other groups (Okada et al. 2005; Okada and Cyster 2006).

5.6 Immunological Synapse and Plasticity of T Cell Activation

The observations of the interaction between T cells and naïve B cells during antigen presentation were rather unexpected (Dustin 2004). This was the first time that long interactions (in the range of several hours) between T cells and APC, which had been postulated in many conceptual frameworks before, were really observed. These interactions were maintained almost exclusively by active by LFA-1, contrary to the dynamic interactions between naïve T cells and DC, which were independent of LFA-1 (Gunzer et al. 2004). However, opposed to the long duration of the contacts, their efficiency for T cell activation was very poor. Thus we asked two questions: which phenotype do T cells develop when they are activated by naïve B cells as APC? And what is the cell-biological mechanism leading to this phenotype?

The most obvious phenotypic difference between T cells activated by naïve B cells and those activated by DC was the stable expression of CD62-L on these cells despite upregulation of CD25 or CD69 and massive proliferation (Reichardt et al. 2007a). T cells activated by DC lost CD62-L rapidly and almost completely, which is the typical early sign of T cell activation (Galkina et al. 2003). The maintenance of high CD62-L levels was obviously mediated by B cells as a slight 'contamination' of the T cell stimulation reaction with 1–10% of DC led to the usual CD62-L loss in the activated T cells. CD62-L on activated T cells was functional, as it mediated preferential homing of the T cells into peripheral lymph nodes of mice after adoptive transfer (Reichardt et al. 2007a).

Also in functional terms B cell-activated T cells were entirely different from their DC-triggered counterparts. We were able to show that T cells activated by B cells inhibited the activation of fresh naïve T cells by DC in vitro. This reaction was titratable for the number of B cell-activated T cells and the cells had to be in physical contact with the naïve target cells in order to mediate suppression (Reichardt et al. 2007a). Strikingly, the suppressive behaviour of the cells was maintained after adoptive transfer into mice. Here the cells could strongly inhibit the priming phase of immune responses. The presence of B cell-induced regulatory T cells inhibited the priming phase of hapten hypersensitivity reactions as well as the rejection of heterotopic allogeneic heart transplants in mice. This worked, however, only when the cells were present before the onset of hapten priming or allograft transplantation, consistent with the preferential homing of the cells into the lymph nodes (Reichardt et al. 2007a). Taken together these results demonstrated, that naïve B cells can indeed be potent APC, but that the main phenotype of activated T cells coming from these reactions has a regulatory potential. These findings might help to explain old observations on the tolerogenic potential of adoptively transferred naïve B cells in vivo (Fuchs and Matzinger 1992).

Thus, naïve T cells have a high level of functional plasticity. The decision about which effector state will be reached has to be made at the earliest possible time point during primary activation. After this decision a T cell line is generated which does not change its phenotype, even after restimulation (Mosmann et al. 1986; Richter et al. 1999). It has been demonstrated that the induction of a stable Th2 phenotype requires the simultaneous triggering of TCR together with the receptor for IL-4 (Richter et al. 1999). Since this happens during the initial interaction with the APC, the functional interaction plane between these two cells, which is termed 'immunological synapse' (IS; (Grakoui et al. 1999; Donnadieu et al. 2001), is of paramount importance for the decision about the final effector phenotype of the activated T cell.

The first IS to be described was between a T cell clone and a B cell line as APC (Monks et al. 1998). It was demonstrated that in the plane of the contact zone a supramolecular sorting of effector and adhesion molecules takes place. In the centre of the interaction plane (central supramolecular activation complex, cSMAC) important signalling molecules are concentrated which are surrounded by a peripheral ring of adhesion molecules (pSMAC). The cSMAC–pSMAC morphology is generated within 30–60 min of onset of an antigen specific T–APC contact (Grakoui et al. 1999). The description of this structure sparked a lot of discussions in the field as it seemed to explain how T cells manage to effectively receive signals despite a relatively low amount of antigen on the APC surface and a low affinity of the TCR towards the cognate pMHC complex (van der Merwe et al. 2000). These signals included the earliest decisions on the type of effector cell which was about to be developed such as the receptor for either IL-4 or IFNγ, under Th2 or Th1 inducing conditions, respectively (Maldonado et al. 2004). It was assumed that T cells would maintain the IS once successfully established and that a cell would permanently receive signals during this phase. Thus it was very surprising when new studies demonstrated that the formation of phosphorylated ZAP70, one of the earliest signals of T cell activation, was terminated before a cSMAC had formed (Lee et al. 2002). It was also striking to see that DC did not form monomeric cSMAC–pSMAC-types of IS but rather multifocal structures (Brossard et al. 2005) and that contacts between T cells and DC, at least in early and late phases of immune responses, were just too short to allow the formation of a proper IS (Mempel et al. 2004). Meanwhile, the function of the IS is being discussed more carefully (Dustin et al. 2005; Mossman et al. 2005). Wet experiments in combination with mathematical modelling showed that the IS is more likely to be associated with regulation of the TCR signalling (Lee et al. 2003) and that signalling starts in the periphery of the IS while the centre is engaged in internalizing triggered TCR (Campi et al. 2005; Yokosuka et al. 2005).

If the decision on the functional phenotype of an activated T cell is taken at the time of IS formation, our experiments described above required that naïve B cells must be forming a regulatory synapse while DC build a classical stimulating IS. Until now no other system has been described that allows such a comparison between functionally entirely different IS. Thus we were very interested to analyse the molecular makeup of these two IS. To our surprise we were able to show that the IS between a naïve B cell and a naïve T cell showed all characteristics of a mature pSMAC–cSMAC-type of IS while DC did not show a comparable organization of molecules in the contact zone with naïve T cells (Fig. 3; Reichardt et al. 2007a).

Ongoing studies shall clarify why a pSMAC–cSMAC type of IS with naïve B cells leads to regulatory T cells. In any case these data demonstrate that the IS is not a monolithic structure but can be very flexible. Probably many different structures of synapses exist which thereby reflect the functional plasticity of effector T cells.

5.7 Migration Control for T Cells Within and out of Lymphatic Organs

In the course of an immune response T cells have several choices after successful activation by DC and clonal expansion. Helper cells may express the chemokine receptor CXCR5 to migrate towards the contact zones of B cell follicles and T zones. Here these T cells help to activate B cells by physical contact as described above (Hardtke et al. 2005). This type of activated T cell is called follicular helper cell (Breitfeld et al. 2000; Hardtke et al. 2005). However, the majority of activated T

Fig. 3. A cSMAC–pSMAC-type of IS forms between naïve T cells and specifically loaded naïve B cells. Contacts of naïve T cells with DC do not show this characteristic sorting of molecules. Naïve DO11.10 T cells, which are specific for a peptide from chicken ovalbumin, were brought into contact with peptide-loaded DC or naïve B cells. Cells were then fixed and the cortical actin cytoskeleton was stained with phalloidin-FITC (*green*) and the TCR with the clonotypic antibody KJ1.26 (*red*). Using a confocal microscope the cells were analysed in different focal planes spaced 0.5 μm apart and spanning the entire cell volume. Shown is a maximal intensity profile in the x–y-orientation of all images from the different focal planes with a DC (**A**) or a naïve B cell (**B**). The Nomarski image of the cell pairs is shown as *a black and white inset*. (**C**) and (**D**) show x–z-projections of the contact zone with a DC or a B cell, respectively. Please note the localization of the entire TCR signal (cSMAC) surrounded by an actin ring (pSMAC) in the T B pair, which cannot be observed in the T–DC pair. Quantification of 300 similar cell pairs emphasizes this point. Modified from research that was originally published in *Blood*. (Reichardt et al. 2007, *Blood*, American Society of Hematology)

cells leave the lymph node in the direction of the medulla where they get back into the blood stream to disperse throughout the entire body. It has been shown that T cells follow a gradient of sphingosine-1-phosphate which they detect by means of their receptor S1P(1) (Matloubian et al. 2004; Schwab et al. 2005). The same mechanism works for the majority of T cells that do not find a cognate DC during their transit through the T zone and therefore are not activated. Recirculating lymphocytes down-regulate their S1P(1) receptor outside of lymph nodes and re-express them again when entering (Lo et al. 2005). Probably, the upregulation of CD69 early after activation of T cells is a mechanism to quickly down-regulate S1P(1) while still within the lymph node. This would inhibit too rapid an exit of naïve T cells from an activating lymph node (Shiow et al. 2006).

5.8 Effector Function in the Periphery

Most activated effector T cells leave the lymph node to exert diverse functions in the periphery of the body (Reinhardt et al. 2001). Also here autonomous migration is essential. The first step of the process of

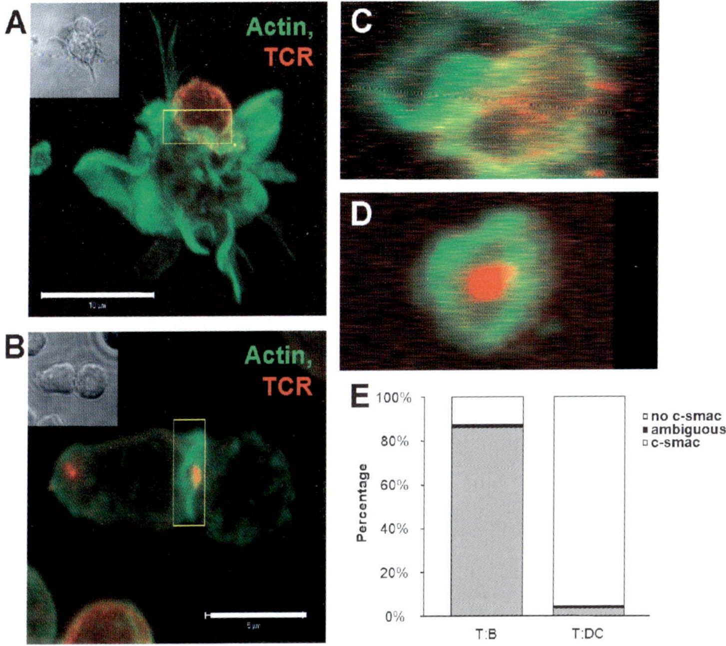

extravasation is the slowing down of cells from the extreme speed in the free blood stream to a slow rolling on the surface of blood vessels (Springer 1994; Berlin et al. 1995). This is only possible at sites where the vessel endothelium expresses E- and P-selectin due to inflammatory reactions (Ebnet and Vestweber 1999). E- and P-selectin are recognized by P-selectin glykoprotein ligand and E-selectin glykoprotein ligand on the surface of effector T cells which are then able to roll on the surface of the endothelium (Ebnet and Vestweber 1999). At sites of higher affinity ligands for integrin-family adhesion receptors such as ICAM-1 or V-CAM are also expressed (Ebnet and Vestweber 1999). This leads to the firm adhesion of T cells, especially via LFA-1 (Weninger et al. 2001). Finally, cells transmigrate between the endothelial cells into the tissue behind. Here they fulfil a variety of effector functions. This may be the activation of local macrophages by means of IFNγ (Mantovani et al. 2004), the destruction of malignant or virus-infected cells (Nestle et al. 1998; Thurner et al. 1999) or the conditioning for the entry of other effector cells (Hwang et al. 2004).

The rate of lymphocytes circulating in the peripheral blood at any given time point is small. In pigs this value is approximately 3% (Pabst and Trepel 1975). From these cells only a small percentage is specific for a given antigen. Nevertheless it is only this peripheral pool that is functionally important for a specific immune response. We could demonstrate this recently by using ligands for Toll like receptor (TLR)7 (Gunzer et al. 2005; Schiller et al. 2006). TLR7 belongs to the group of TLR which recognize evolutionarily highly conserved pathogen-associated molecular patterns (PAMPs; Takeda et al. 2003). Binding of such a PAMP to TLR leads to a complex cascade of signalling events in the triggered cell. At the end of this process many effector functions such as increased migration, proliferation or release of proinflammatory cytokines are switched on (Akira and Takeda 2004). The natural PAMP for TLR7 is viral single-stranded RNA (Diebold et al. 2004; Heil et al. 2004). We tested the effect of a systemic dose of R848, which is an artificial ligand for TLR7 (Gunzer et al. 2005). While this drug had strongly stimulating effects on hapten-dependent contact hypersensitivity (CHS) reactions in a time course of 6 days it also induced a rapid and almost complete leukopenia within min of application which lasted for 36 h. This allowed us to test how important the pool of circulating effector cells in the blood is for the elicitation of peripheral T cell mediated effector functions.

Indeed, the depletion of the peripheral pool led to an almost complete suppression of the hapten-mediated CHS response in experimental animals. Only after reappearance of the peripheral cells could the reaction be elicited again (Gunzer et al. 2005). This shows prominently that a relatively small number of T cells is sufficient to mount a specific immune response, if the cells are available at the right place. In reverse it means that even large numbers of specific effector T cells cannot exert their function if they are wrongly positioned. A novel immunosuppressive drug, FTY720, exploits just this principle. Via stimulation of S1P(1) it down-regulates this receptor on leukocytes throughout the body and also on cells within lymph nodes. Now these cells are blind for the gradient of S1P leading them out of lymphatic tissue and are thus trapped in lymph nodes (Chiba et al. 1998; Quesniaux et al. 1999). FTY720 treated animals or patients are thus strongly immunosuppressed. It is also not surprising that such mechanisms are being used by pathogens (Zinkernagel 1996). Many viral infections lead to leukopenia (Tumpey et al. 2000). The mechanisms underlying these effects are not well understood. It needs to be tested whether the stimulation, for example of TLR7 by viral single-stranded RNA is such a mechanism. This might possibly explain why patients suffering from a viral infection are extremely susceptible to a bacterial superinfection (Okamoto et al. 2003; Peltola and McCullers 2004).

6 Migration and Cellular Dynamics of the Innate Immune System

The previous part of this review dealt exclusively with the migration processes associated with the generation and function of the adaptive immune system. However, one must not forget that when measured in cell numbers and cellular turnover the innate immune system constitutes by far the largest part of cellular immunity. It is also important to note that the missing or decline of adaptive immunity leads to health problems only gradually, sometimes over a period of years, as prominently demonstrated in the case of AIDS (Kaufmann and McMichael 2005). In sharp contrast, the state of neutropenia, which means the absence or too low a concentration of neutrophil granulocytes (polymorphonuclear cells, PMN) in an individual, is often acutely fatal as no defence against normally harmless infections is possible. This can be demonstrated for the infection with the ubiquitous environmental mold *Aspergillus fumigatus* (Latge 2001). The spores of this fungus are present in normal breathing air and are inhaled by us every day without the occurrence of health problems. The same fungus, however, can cause barely treatable, fatal invasive diseases in patients who suffer from a therapy-induced neutropenia or a genetic defect in PMN function such as chronic granulomatous disease (Latge 1999, 2001). Thus, the analysis of innate immune cells and, in the context of this review, an in depth knowledge about their migration is indispensable to get a complete picture of the importance of dynamic cellular immunological processes.

6.1 Recruitment into Inflamed Regions

A central step in the function of the innate immune system is the recruitment of effector cells to the place of function. Already Ilya Metchnikoff had observed the recruitment of phagocytes in sea urchins to spots where he had introduced a prick (Tauber 2003; Metchnikoff 1883). It was also Metchnikoff who coined the terms ‘phagocyte’ as well as ‘macrophage’. Consequently he also invented the ‘microphage’, the only character that did not survive the times. Today, Metchnikoffs ‘microphages’ are called PMN (Tauber 2003).

PMN, the main cell type of early inflammatory processes are mobilized from bone marrow precursors into the peripheral blood and from there recruited into inflamed tissue sites (Nathan 2006; Christopher and Link 2007). Thereby blood borne PMN follow the same rules of rolling, firm adhesion and diapedesis as described above for T cells (Springer 1994). Within the tissue PMN are able to migrate in a highly directed fashion towards inflammatory foci. This process is called chemotaxis because PMN are able to sense gradients of chemokines secreted by body cells which are already located in the inflamed region. IL-8 is a prominent example of such a PMN-attractive chemokine which is produced by tissue resident macrophages (Sozzani et al. 1995; Moghe et al. 1995) and triggers the motility of PMN by means of calcium-binding proteins of the S100 family (Manitz et al. 2003). There is a large number of other chemokines which can be endogenously produced. For example, it has been shown that during infection with *A. fumigatus* the chemokine receptor CXCR2 is important as it allows cells to migrate along gradients of the chemokines MIP-2 or KC (Bonnett et al. 2006). In addition to their ability to migrate towards sources of endogenous chemokines PMN also have receptors to directly sense pathogens or their breakdown products. A particularly effective system is a receptor which recognizes bacterial proteins by means of their specific signature, the presence of formylated methionine residues (Verploegen et al. 2005). Bacterial proteins normally start with a formylated methionine residue, while eukaryotic proteins do not. Thus, the presence of fMLP (formyl-Methionyl-Leucyl-Phenylalanine) is recognized by PMN with their fMLP receptor (Snyderman and Pike 1984).

Measurements of PMN chemotaxis usually focus on the counting of transmigrated cells in Boyden chambers (Boyden 1962), which is called an indirect method, as the cells are not observed during migration. Only very few studies measure cell migration directly (e.g. Moghe et al. 1995). However, important phenomena such as chemokinesis, which is the generally increased migration in the absence of a main orientation, often remain undetected when using indirect methods. Thus we have analysed the migration of human PMN to a source of fMLP within an artificial three-dimensional matrix (Entschladen et al. 2000). Here we could demonstrate that only during the first 15 min after onset of an fMLP gradient there is directed migration. However, massive chemokinesis induced by fMLP was detectable for more that 2 h afterwards (Entschladen et al. 2000). The increase of chemokinesis was very profound—within seconds of wash-in of the compound almost all cells reacted. From a basal level of 10%–20% migratory activity, up to 70%–90% of cells were induced to migrate. In parallel the migration velocity was increased 300%. In contrast, T cells migrating in response to a gradient of SDF-1 did this at the same velocity. SDF recruited only previously immobile cells (Entschladen et al. 2000). This paper showed for the first time that distinct phases of chemotaxis and chemokinesis can be induced by the same bacterial chemotaktin and can last for very different scales of time. Also other receptors on PMN such as the proteinase activated receptor 2 (PAR2), which can be induced by bacterial proteases, induce strong chemokinesis after triggering in three-dimensional collagen matrices (Shpacovitch et al. 2004). Conceptionally, chemokinesis may be as important for bacterial destruction as chemotaxis because at drastically increased PMN velocities their chances of encountering bacteria for phagocytosis are also largely increased.

Increased motility allows PMN to enter inflamed sites rapidly. Within 6 h after inhalation of *A. fumigatus* conidia the number of PMN in the lung is increased by a factor of 500 while they represent only trace populations in non-infected lungs (Bonnett et al. 2006). It is likely that PMN also show increased motility after immigration into inflamed sites, such as the lung. This should be analysed by intravital microscopy in the future.

6.2 Effector Functions of PMN

After immigration into inflamed sites PMN perform two main functions: (1) production and release of aggressive agents such as proteases and oxygen radicals for extracellular destruction of pathogens, and (2) the phagocytosis of pathogens followed by intracellular digestion (Nathan 2006; Segal 2005). While the relevant cell-biological mechanisms are well defined, the underlying cellular dynamics are incompletely understood. Thus, we have recently analysed the interaction of PMN and alveolar macrophages or DC with conidia of *A. fumigatus*. In this study we analysed the enormous cellular dynamics of phagocytosis by different phagocytes (Behnsen et al. 2007). We also demonstrated a novel type of interaction between PMN and *A. fumigatus* conidia, which we have termed dragging. Here, conidia are not phagocytosed but collected on the surface of PMN in large numbers (Fig. 4).

PMN loaded with fungal conidia are still migrating and come together in large clusters centred by fungal spores (Behnsen et al. 2007). Such clusters are also formed in vivo as shown recently in mice (Bonnett et al. 2006). Probably they are important for the destruction of large numbers of conidia outside of PMN in a controlled fashion as in these clusters the generation of reactive oxygen species is focused to the zones where PMN directly contact the conidial surface (Bonnett et al. 2006).

In the same study we made another unexpected observation. While PMN and alveolar macrophages phagocytosed conidia of *A. fumigatus* extremely efficiently in two-dimensional liquid medium cultures they were unable to do this in three-dimensional collagen matrices. This was not due to a general inability of PMN to phagocytose fungal elements in three-dimensional environments because the same cells efficiently phagocytosed spores of *Candida albicans* in three- but not in two-dimensional environments. In competitive phagocytosis experiments, where equal numbers of both fungal pathogens were available either in two or three-dimensional environments PMN selectively incorporated *Aspergillus* conidia in two-dimensional and *Candida* spores in three-dimensional environments, despite frequent physical contacts with both pathogens (Behnsen et al. 2007).

We interpret these results as a suggestion that phagocytes can adapt their functionality to the environment. On the surface of alveoli the

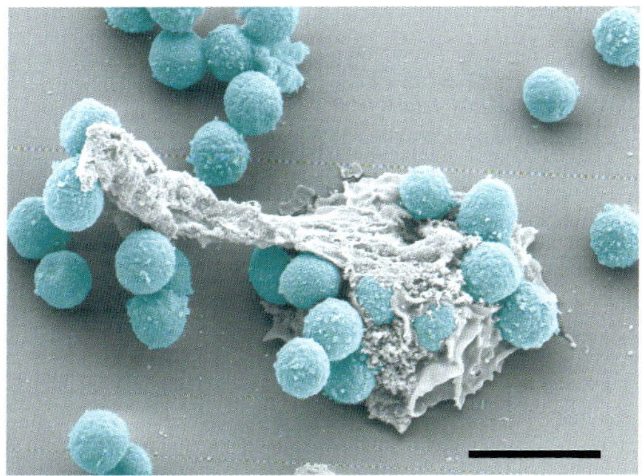

Fig. 4. A cluster of a PMN and dragged as well as phagocytosed conidia of *A. fumigatus*. Murine bone marrow PMN were mixed with conidia of *A. fumigatus*, incubated for 30 min and then fixed for analysis by scanning electron microscopy. Note the occurrence of loosely attached ('dragged') conidia as well as conidia, which are already deeply embedded in the membrane protrusions of the PMN. This is a sign of ongoing phagocytosis. In addition the PMN extends a membrane protrusion to the left to catch nearby conidia. (Dynamic images from such interactions can be downloaded from the supplemental material of Behnsen et al. 2007). Scalebar: 5 μm

lung really is a two-dimensional environment. This surface is formed by type I epithelial cells; lung phagocytes are lying on this surface and are protected from direct contact with the breathing air by a thin film of liquid, the surface lining layer (SLL). The SLL consists of a mixture of water and surfactant (itself a mixture of lipoproteins and phospholipids) and is produced by type II epithelial cells of the lung (Gehr et al. 1996; Geiser and Bastian 2003). Ultrastructural analyses of the SLL show that the phagocytes beneath it are very flat and pressed against the surface formed by the type I epithelial cells (Gehr et al. 1996; Geiser 2002). The SLL is very thin (~50 nm) and therefore has no room for migration of cells into space. Inhaled particles, like fungal spores, are drawn beneath the SLL and then internalized by the phagocytes of the lung (Geiser 2002). Defects in the surfactant system, for example in newborns, lead to the so-called respiratory distress syndrome. Here the lung collapses entirely during each exhalation and needs to be re-inflated at each inhalation with massive force (Tortora and Grabowski 2000). A functional surfactant system reduces the surface tension of the SLL-watery phase so much that the collapse of alveoli during exhalation is inhibited. Thus, the alveolus of a normally functioning lung is a true two-dimensional environment for phagocytes working within. Normally, PMN meet with inhaled conidia of *A. fumigatus* only in this zone of the body. Since this mould is very common in the environment and has accompanied humans during their evolution, it is conceivable that the phagocytosis of lung-borne cells have been optimized for this special two-dimensional environment. In contrast, *C. albicans* is a commensal fungus of human mucosa. Only in the context of immunological dysfunction can the fungus grow invasively into tissues. Then it will be attacked by PMN within the three-dimensional environment of tissue microabscesses (Newman et al. 2005). As this phenomenon is not ‘new’ in evolutionary terms, a functional adaptation of phagocyte and pathogen to the environment, where encounter is most likely to occur, would be imaginable.

Thus, future work needs to clarify the molecular mechanisms behind these striking differences in phagocytic activity of PMN. It is possible that the pathogens change their surface composition when coming from the ‘natural’ to the ‘unusual’ environment (e.g., the conidia of *A. fumigatus* during transition from the two-dimensional environment of the lung to the three-dimensional environment of tissues) in a way to make them ‘invisible’ to phagocytes. But phagocytes might sense the dimensionality of the environment and react to this with a change in adhesion receptors that no longer allows interaction with the pathogen, that is naturally encountered in another environment. This might finally explain why infections with a pathogen at unusual sites can be very dangerous. Healthy mice easily survive lung infections with up to 2×10^8 *A. fumigatus* conidia but succumb to 40 times fewer conidia when they are given intravenously (Garlanda et al. 2002).

7 Conclusion

The analysis of cellular dynamics and migration is of paramount importance for a complete understanding of normal immune function and its failure in the state of disease. The past 5 years have seen an enormous increase in knowledge about these phenomena, based primarily on the advent of new intravital imaging approaches. In the future this knowledge will increase at accelerated speed as more and more groups learn how to adapt the technology to their specific questions. Thus, exciting times in cellular immunology lie ahead and the interested reader should keep an eye on this area of immunological research.

References

Ackerman AL, Cresswell P (2004) Cellular mechanisms governing cross-presentation of exogenous antigens. Nat Immunol 5:678–684

Adams GB, Chabner KT, Alley IR, Olson DP, Szczepiorkowski ZM, Poznansky MC, Kos CH, Pollak MR, Brown EM, Scadden DT (2006) Stem cell engraftment at the endosteal niche is specified by the calcium-sensing receptor. Nature 439:599–603

Akira S, Takeda K (2004) Toll-like receptor signalling. Nat Rev Immunol 4:499–511

Allen CD, Okada T, Tang HL, Cyster JG (2007) Imaging of germinal center selection events during affinity maturation. Science 315:528–531

Aydar Y, Sukumar S, Szakal AK, Tew JG (2005) The influence of immune complex-bearing follicular dendritic cells on the IgM response, Ig class switching, and production of high affinity IgG. J Immunol 174:5358–5366

Babbitt BP, Allen PM, Matsueda G, Haber E, Unanue ER (1985) Binding of immunogenic peptides to Ia histocompatibility molecules. Nature 317:359–361

Banchereau J, Bazan F, Blanchard D, Briere F, Galizzi JP, van Kooten C, Liu YJ, Rousset F, Saeland S (1994) The CD40 antigen and its ligand. Annu Rev Immunol 12:881–922

Banchereau J, Briere F, Caux C, Davoust J, Lebecque S, Liu YJ, Pulendran B, Palucka K (2000) Immunobiology of dendritic cells. Annu Rev Immunol 18:767–811

Banchereau J, Palucka AK (2005) Dendritic cells as therapeutic vaccines against cancer. Nat Rev Immunol 5:296–306

Behnsen J, Narang P, Hasenberg M, Gunzer F, Bilitewski U, Klippel N, Rohde M, Brock M, Brakhage AA, Gunzer M (2007) Environmental dimensionality controls the interaction of phagocytes with the pathogenic fungi *Aspergillus fumigatus* and *Candida albicans*. PLoS Pathog 3:e13

Benham A, Tulp A, Neefjes J (1995) Synthesis and assembly of MHC-peptide-complexes. Immunol Today 16:359–362

Berlin C, Bargatze RF, Campbell JJ, von Andrian UH, Szabo MC, Hasslen SR, Nelson RD, Berg EL, Erlandsen SL, Butcher EC (1995) α4 Integrins mediate lymphocyte attachment and rolling under physiologic flow. Cell 80:413–422

Bhakta NR, Oh DY, Lewis RS (2005) Calcium oscillations regulate thymocyte motility during positive selection in the three-dimensional thymic environment. Nat Immunol 6:143–151

Bjorkman PJ, Saper MA, Samraoui B, Bennett WS, Strominger JL, Wiley DC (1987a) Structure of the human class I histocompatibility antigen, HLA-A2. Nature 329:506–512

Bjorkman PJ, Saper MA, Samraoui B, Bennett WS, Strominger JL, Wiley DC (1987b) The foreign antigen binding site and T cell recognition regions of class I histocompatibility antigens. Nature 329:512–518

Boehm T, Bleul CC (2006) Thymus-homing precursors and the thymic microenvironment. Trends Immunol 27:477–484

Bonnett CR, Cornish EJ, Harmsen AG, Burritt JB (2006) Early neutrophil recruitment and aggregation in the murine lung inhibit germination of *Aspergillus fumigatus* conidia. Infect Immun 74:6528–6539

Bousso P, Bhakta NR, Lewis RS, Robey E (2002) Dynamics of thymocyte-stromal cell interactions visualized by two-photon microscopy. Science 296:1876–1880

Bousso P, Robey E (2003) Dynamics of CD8(+) T cell priming by dendritic cells in intact lymph nodes. Nat Immunol 4:579–585

Bousso P, Robey EA (2004) Dynamic behavior of T cells and thymocytes in lymphoid organs as revealed by two-photon microscopy. Immunity 21:349–355

Boyden S (1962) The chemotactic effect of mixtures of antibody and antigen on polymorphonuclear leucocytes. J Exp Med 115:453–466

Breitfeld D, Ohl L, Kremmer E, Ellwart J, Sallusto F, Lipp M, Forster R (2000) Follicular B helper T cells express CXC chemokine receptor 5, localize to B cell follicles, and support immunoglobulin production. J Exp Med 192:1545–1552

Brossard C, Feuillet V, Schmitt A, Randriamampita C, Romao M, Raposo G, Trautmann A (2005) Multifocal structure of the T cell—dendritic cell synapse. Eur J Immunol 35:1741–1753

Cahalan MD, Parker I, Wei SH, Miller MJ (2002) Two-photon tissue imaging: seeing the immune system in a fresh light. Nat Rev Immunol 2:872–880

Campi G, Varma R, Dustin ML (2005) Actin and agonist MHC-peptide complex-dependent T cell receptor microclusters as scaffolds for signaling. J Exp Med 202:1031–1036

Castellino F, Huang AY, tan-Bonnet G, Stoll S, Scheinecker C, Germain RN (2006) Chemokines enhance immunity by guiding naive $CD8^+$ T cells to sites of CD4+ T cell-dendritic cell interaction. Nature 440:890–895

Chiba K, Yanagawa Y, Masubuchi Y, Kataoka H, Kawaguchi T, Ohtsuki M, Hoshino Y (1998) FTY720, a novel immunosuppressant, induces sequestration of circulating mature lymphocytes by acceleration of lymphocyte homing in rats. I. FTY720 selectively decreases the number of circulating mature lymphocytes by acceleration of lymphocyte homing. J Immunol 160:5037–5044

Chow AY, Mellman I (2005) Old lysosomes, new tricks: MHC II dynamics in DCs. Trends Immunol 26:72–78

Christopher MJ, Link DC (2007) Regulation of neutrophil homeostasis. Curr Opin Hematol 14:3–8

Crawford A, Macleod M, Schumacher T, Corlett L, Gray D (2006) Primary T cell expansion and differentiation in vivo requires antigen presentation by B cells. J Immunol 176:3498–3506

Cyster JG (1999) Chemokines and cell migration in secondary lymphoid organs. Science 286:2098–2102

de Vinuesa CG, Cook MC, Ball J, Drew M, Sunners Y, Cascalho M, Wabl M, Klaus GG, MacLennan IC (2000) Germinal centers without T cells. J Exp Med 191:485–494

Diebold SS, Kaisho T, Hemmi H, Akira S, Reis e Sousa C (2004) Innate antiviral responses by means of TLR7-mediated recognition of single-stranded RNA. Science 303:1529–1531

Donnadieu E, Revy P, Trautmann A (2001) Imaging T-cell antigen recognition and comparing immunological and neuronal synapses. Immunology 103:417–425

Dorner B, Müller S, Entschladen F, Schröder JM, Franke P, Kraft R, Friedl P, Clark-Lewis I, Kroczek RA (1997) Purification, structural analysis, and function of ATAC, a cytokine secreted by $CD8^+$ T cells. J Biol Chem 272:8817–8823

Dustin ML (2004) New ways for lyphocytes to meet. Blood 104:2618–2619

Dustin ML (2005) A dynamic view of the immunological synapse. Semin Immunol 17:400–410

Dustin ML, Allen PM, Shaw AS (2001) Environmental control of immunological synapse formation and duration. Trends Immunol 22:192–194

Dustin ML, Chan AC (2000) Signaling takes shape in the immune system. Cell 103:283–294

Dustin ML, de Fougerolles AR (2001) Reprograming T cells: the role of extracellular matrix in coordination of T cell activation and migration. Curr Opin Immunol 13:286–290

Ebnet K, Vestweber D (1999) Molecular mechanisms that control leukocyte extravasation: the selectins and the chemokines. Histochem Cell Biol 112:1–23

Entschladen F, Gunzer M, Scheuffele CM, Niggemann B, Zänker KS (2000) T lymphocytes and neutrophil granulocytes differ in regulatory signaling and migratory dynamics with regard to spontaneous locomotion and chemotaxis. Cell Immunol 199:104–114

Entschladen F, Niggemann B, Zänker KS, Friedl P (1997) Differential requirement of protein tyrosine kinases and protein kinase C in the regulation of T cell locomotion in three-dimensional collagen matrices. J Immunol 159:3203–3210

Epstein MM, Di RF, Jankovic D, Sher A, Matzinger P (1995) Successful T cell priming in B cell-deficient mice. J Exp Med 182:915–922

Fillatreau S, Sweenie CH, McGeachy MJ, Gray D, Anderton SM (2002) B cells regulate autoimmunity by provision of IL-10. Nat Immunol 3:944–950

Förster R, Mattis AE, Kremmer E, Wolf E, Brem G, Lipp M (1996) A putative chemokine receptor, BLR1, directs B cell migration to defined lymphoid organs and specific anatomic compartments of the spleen. Cell 87:1037–1047

Förster R, Schubel A, Breitfeld D, Kremmer E, Renner-Müller I, Wolf E, Lipp M (1999) CCR7 coordinates the primary immune response by establishing functional microenvironments in secondary lymphoid organs. Cell 99:23–33

Friedl P, Entschladen F, Conrad C, Niggemann B, Zänker KS (1998) CD4+ T lymphocytes migrating in three-dimensional collagen lattices lack focal adhesions and utilize beta1 integrin-independent strategies for polarization, interaction with collagen fibers and locomotion. Eur J Immunol 28:2331–2343

Friedl P, Gunzer M (2001) Interaction of T cells with APCs: the serial encounter model. Trends Immunol 22:187–191

Friedl P, Noble PB, Shields ED, Zänker KS (1994) Locomotor phenotypes of unstimulated CD45RAhigh and CD45ROhigh CD4+ and CD8+ lymphocytes in three-dimensional collagen lattices. Immunology 82:617–624

Friedl P, Noble PB, Zänker KS (1993) Lymphocyte migration in three-dimensional collagen gels. Comparison of three quantitative methods for analysing cell trajectories. J Immunol Meth 165:157–165

Friedl P, Noble PB, Zänker KS (1995) T Lymphocyte locomotion in a three-dimensional collagen matrix. Expression and function of cell adhesion molecules. J Immunol 154:4973–4985

Fuchs EJ, Matzinger P (1992) B cells turn off virgin but not memory T cells. Science 258:1156–1159

Galkina E, Tanousis K, Preece G, Tolaini M, Kioussis D, Florey O, Haskard DO, Tedder TF, Ager A (2003) L-selectin shedding does not regulate constitutive T cell trafficking but controls the migration pathways of antigen activated T lymphocytes. J Exp Med 198:1323–1335

Garcia Z, Pradelli E, Celli S, Beuneu H, Simon A, Bousso P (2007) Competition for antigen determines the stability of T cell-dendritic cell interactions during clonal expansion. Proc Natl Acad Sci USA 104:4553–4558

Garlanda C, Hirsch E, Bozza S, Salustri A, De Acetis M, Nota R, Maccagno A, Riva F, Bottazzi B, Peri G, Doni A, Vago L, Botto M, De Santis R, Carminati P, Siracusa G, Altruda F, Vecchi A, Romani L, Mantovani A (2002) Non-redundant role of the long pentraxin PTX3 in anti-fungal innate immune response. Nature 420:182–186

Garside P, Ingulli E, Merica RR, Johnson JG, Noelle RJ, Jenkins MK (1998) Visualization of specific B and T lymphocyte interactions in the lymph node. Science 281:96–99

Gehr P, Green FH, Geiser M, Im HV, Lee MM, Schurch S (1996) Airway surfactant, a primary defense barrier: mechanical and immunological aspects. J Aerosol Med 9:163–181

Geiser M (2002) Morphological aspects of particle uptake by lung phagocytes. Microsc Res Tech 57:512–522

Geiser M, Bastian S (2003) Surface-lining layer of airways in cystic fibrosis mice. Am J Physiol Lung Cell Mol Physiol 285:L1277–L1285

Germain RN (2002) T-cell development and the CD4-CD8 lineage decision. Nat Rev Immunol 2:309–322

Germain RN, Jenkins MK (2004) In vivo antigen presentation. Curr Opin Immunol 16:120–125

Godin I, Cumano A (2005) Of birds and mice: hematopoietic stem cell development. Int J Dev Biol 49:251–257

Grakoui A, Bromley SK, Sumen C, Davis MM, Shaw AS, Allen PM, Dustin ML (1999) The immunological synapse: a molecular machine controlling T cell activation. Science 285:221–227

Gunzer M, Friedl P, Niggemann B, Bröcker E-B, Kämpgen E, Zänker KS (2000a) Migration of dendritic cells within 3-D collagen lattices is dependent on tissue origin, state of maturation, and matrix structure and is maintained by proinflammatory cytokines. J Leukoc Biol 67:622–629

Gunzer M, Janich S, Varga G, Grabbe S (2001) Dendritic cells and tumor immunity. Semin Immunol 13:291–302

Gunzer M, Kämpgen E, Bröcker E-B, Zänker KS, Friedl P (1997) Migration of dendritic cells in 3D-collagen lattices: visualisation of dynamic interactions with the substratum and the distribution of surface structures via a novel confocal reflection imaging technique. Adv Exp Med Biol 417:97–103

Gunzer M, Riemann H, Basoglu Y, Hillmer A, Weishaupt C, Balkow S, Benninghoff B, Ernst B, Steinert M, Scholzen T, Sunderkotter C, Grabbe S (2005) Systemic administration of a TLR7 ligand leads to transient immune incompetence due to peripheral blood leukocyte depletion. Blood 106:2424–2432

Gunzer M, Schäfer A, Borgmann S, Grabbe S, Zänker KS, Bröcker E-B, Kämpgen E, Friedl P (2000b) Antigen presentation in extracellular matrix: interactions of T cells with dendritic cells are dynamic, short lived, and sequential. Immunity 13:323–332

Gunzer M, Weishaupt C, Hillmer A, Basoglu Y, Friedl P, Dittmar KE, Kolanus W, Varga G, Grabbe S (2004) A spectrum of biophysical interaction modes between T cells and different antigen presenting cells during priming in 3-D collagen and in vivo. Blood 104:2801–2809

Hardtke S, Ohl L, Forster R (2005) Balanced expression of CXCR5 and CCR7 on follicular T helper cells determines their transient positioning to lymph node follicles and is essential for efficient B-cell help. Blood 106:1924–1931

Harrington LE, Hatton RD, Mangan PR, Turner H, Murphy TL, Murphy KM, Weaver CT (2005) Interleukin 17-producing CD4(+) effector T cells develop via a lineage distinct from the T helper type 1 and 2 lineages. Nat Immunol 6:1123–1132

Heil F, Hemmi H, Hochrein H, Ampenberger F, Kirschning C, Akira S, Lipford G, Wagner H, Bauer S (2004) Species-specific recognition of single-stranded RNA via Toll-like receptor 7 and 8. Science 303:1526–1529

Huang NN, Han SB, Hwang IY, Kehrl JH (2005) B cells productively engage soluble antigen-pulsed dendritic cells: visualization of live-cell dynamics of B cell-dendritic cell interactions. J Immunol 175:7125–7134

Hugues S, Fetler L, Bonifaz L, Helft J, Amblard F, Amigorena S (2004) Distinct T cell dynamics in lymph nodes during the induction of tolerance and immunity. Nat Immunol 5:1235–1242

Hwang JM, Yamanouchi J, Santamaria P, Kubes P (2004) A critical temporal window for selectin-dependent CD4+ lymphocyte homing and initiation of late-phase inflammation in contact sensitivity. J Exp Med 199:1223–1234

Iezzi G, Karjalainen K, Lanzavecchia A (1998) The duration of antigenic stimulation determines the fate of naive and effector T cells. Immunity 8:89–95

Kaufmann SH, McMichael AJ (2005) Annulling a dangerous liaison: vaccination strategies against AIDS and tuberculosis. Nat Med 11:S33–S44

Knoechel B, Lohr J, Kahn E, Abbas AK (2005) Cutting edge: the link between lymphocyte deficiency and autoimmunity: roles of endogenous T and B lymphocytes in tolerance. J Immunol 175:21–26

Kosco-Vilbois MH (2003) Opinion: are follicular dendritic cells really good for nothing? Nat Rev Immunol 3:764–769
Lanzavecchia A, Iezzi G, Viola A (1999) From TCR engagement to T cell activation: a kinetic view of T cell behavior. Cell 96:1–4
Lanzavecchia A, Sallusto F (2001a) Antigen decoding by T lymphocytes: from synapses to fate determination. Nat Immunol 2:487–492
Lanzavecchia A, Sallusto F (2001b) The instructive role of dendritic cells on T cell responses: lineages, plasticity and kinetics. Curr Opin Immunol 13:291–298
Lassila O, Vainio O, Matzinger P (1988) Can B cells turn on virgin T cells? Nature 334:253–255
Latge JP (1999) *Aspergillus fumigatus* and aspergillosis. Clin Microbiol Rev 12:310–350
Latge JP (2001) The pathobiology of *Aspergillus fumigatus*. Trends Microbiol 9:382–389
Lee KH, Dinner AR, Tu C, Campi G, Raychaudhuri S, Varma R, Sims TN, Burack WR, Wu H, Wang J, Kanagawa O, Markiewicz M, Allen PM, Dustin ML, Chakraborty AK, Shaw AS (2003) The immunological synapse balances T cell receptor signaling and degradation. Science 302:1218–1222
Lee KH, Holdorf AD, Dustin ML, Chan AC, Allen PM, Shaw AS (2002) T cell receptor signaling precedes immunological synapse formation. Science 295:1539–1542
Lindquist RL, Shakhar G, Dudziak D, Wardemann H, Eisenreich T, Dustin ML, Nussenzweig MC (2004) Visualizing dendritic cell networks in vivo. Nat Immunol 5:1243–1250
Lipsky PE, Rosenthal AS (1975) Macrophage-lymphocyte interaction. II. Antigen-mediated physical interactions between immune guinea pig lymph node lymphocytes and syngeneic macrophages. J Exp Med 141:138–154
Lo CG, Xu Y, Proia RL, Cyster JG (2005) Cyclical modulation of sphingosine-1-phosphate receptor 1 surface expression during lymphocyte recirculation and relationship to lymphoid organ transit. J Exp Med 201:291–301
MacLennan IC (1994) Germinal centers. Annu Rev Immunol 12:117–139
Maldonado RA, Irvine DJ, Schreiber R, Glimcher LH (2004) A role for the immunological synapse in lineage commitment of CD4 lymphocytes. Nature 431:527–532
Manitz MP, Horst B, Seeliger S, Strey A, Skryabin BV, Gunzer M, Frings W, Schonlau F, Roth J, Sorg C, Nacken W (2003) Loss of S100A9 (MRP14) results in reduced interleukin-8-induced CD11b surface expression, a polarized microfilament system, and diminished responsiveness to chemoattractants in vitro. Mol Cell Biol 23:1034–1043
Manser T (2004) Textbook germinal centers? J Immunol 172:3369–3375

Mantovani A, Sica A, Sozzani S, Allavena P, Vecchi A, Locati M (2004) The chemokine system in diverse forms of macrophage activation and polarization. Trends Immunol 25:677–686

Matloubian M, Lo CG, Cinamon G, Lesneski MJ, Xu Y, Brinkmann V, Allende ML, Proia RL, Cyster JG (2004) Lymphocyte egress from thymus and peripheral lymphoid organs is dependent on S1P receptor 1. Nature 427:355–360

Matthias P, Rolink AG (2005) Transcriptional networks in developing and mature B cells. Nat Rev Immunol 5:497–508

Matzinger P (1994) Tolerance, danger, and the extended family. Annu Rev Immunol 12:991–1045

Mempel TR, Henrickson SE, von Andrian UH (2004) T-cell priming by dendritic cells in lymph nodes occurs in three distinct phases. Nature 427:154–159

Mempel TR, Pittet MJ, Khazaie K, Weninger W, Weissleder R, von Boehmer H, von Andrian UH (2006) Regulatory T cells reversibly suppress cytotoxic T cell function independent of effector differentiation. Immunity 25:129–141

Metchnikoff II (1883) Untersuchungen über die Mesodermalen Phagozyten einiger Wirbeltiere. Biol Zent Bl 3:560–565

Meyer-Hermann ME, Maini PK (2005) Cutting edge: back to 'one-way' germinal centers. J Immunol 174:2489–2493

Miller MJ, Hejazi AS, Wei SH, Cahalan MD, Parker I (2004a) T cell repertoire scanning is promoted by dynamic dendritic cell behavior and random T cell motility in the lymph node. Proc Natl Acad Sci USA 101:998–1003

Miller MJ, Safrina O, Parker I, Cahalan MD (2004b) Imaging the Single Cell Dynamics of CD4+ T Cell Activation by Dendritic Cells in Lymph Nodes. J Exp Med 200:847–856

Miller MJ, Wei SH, Parker I, Cahalan MD (2002) Two-photon imaging of lymphocyte motility and antigen response in intact lymph node. Science 296:1869–1873

Misslitz A, Pabst O, Hintzen G, Ohl L, Kremmer E, Petrie HT, Forster R (2004) Thymic T Cell Development and Progenitor Localization Depend on CCR7. J Exp Med 200:481–491

Mizoguchi A, Bhan AK (2006) A case for regulatory B cells. J Immunol 176:705–710

Moghe PV, Nelson RD, Tranquillo RT (1995) Cytokine-stimulated chemotaxis of human neutrophils in a 3-D conjoined fibrin gel assay. J Immunol Methods 180:193–211

Monks CR, Freiberg BA, Kupfer H, Sciaky N, Kupfer A (1998) Three-dimensional segregation of supramolecular activation clusters in T cells. Nature 395:82–86

Moser K, Tokoyoda K, Radbruch A, MacLennan I, Manz RA (2006) Stromal niches, plasma cell differentiation and survival. Curr Opin Immunol 18:265–270

Mosmann TR, Cherwinski H, Bond MW, Giedlin MA, Coffman RL (1986) Two types of murine helper T cell clone. I. Definition according to profiles of lymphokine activities and secreted proteins. J Immunol 136:2348–2357

Mossman KD, Campi G, Groves JT, Dustin ML (2005) Altered TCR signaling from geometrically repatterned immunological synapses. Science 310:1191–1193

Murphy KM, Heimberger AB, Loh DY (1990) Induction by antigen of intrathymic apoptosis of $CD4^{+}CD8^{+}TCR^{lo}$ thymocytes in vivo. Science 250:1720–1723

Nathan C (2006) Neutrophils and immunity: challenges and opportunities. Nat Rev Immunol 6:173–182

Nestle FO, Alijagic S, Gilliet M, Sun Y, Grabbe S, Dummer R, Burg G, Schadendorf D (1998) Vaccination of melanoma patients with peptide – or tumor lysate – pulsed dendritic cells [see comments]. Nat Med 4:328–332

Newman SL, Bhugra B, Holly A, Morris RE (2005) Enhanced killing of *Candida albicans* by human macrophages adherent to type 1 collagen matrices via induction of phagolysosomal fusion. Infect Immun 73:770–777

Norbury CC, Basta S, Donohue KB, Tscharke DC, Princiotta MF, Berglund P, Gibbs J, Bennink JR, Yewdell JW (2004) CD8+ T cell cross-priming via transfer of proteasome substrates. Science 304:1318–1321

Okada T, Cyster JG (2006) B cell migration and interactions in the early phase of antibody responses. Curr Opin Immunol 18:278–285

Okada T, Miller MJ, Parker I, Krummel MF, Neighbors M, Hartley SB, O'Garra A, Cahalan MD, Cyster JG (2005) Antigen-engaged B cells undergo chemotaxis toward the T zone and form motile conjugates with helper T cells. PLoS Biol 3:e150

Okamoto S, Kawabata S, Nakagawa I, Okuno Y, Goto T, Sano K, Hamada S (2003) Influenza A virus-infected hosts boost an invasive type of Streptococcus pyogenes infection in mice. J Virol 77:4104–4112

Pabst R, Trepel F (1975) Quantitative evaluation of the total number and distribution of lymphocytes in young pigs. Blut 31:77–86

Peltola VT, McCullers JA (2004) Respiratory viruses predisposing to bacterial infections: role of neuraminidase. Pediatr Infect Dis J 23:S87–S97

Qi H, Egen JG, Huang AY, Germain RN (2006) Extrafollicular activation of lymph node B cells by antigen-bearing dendritic cells. Science 312:1672–1676

Quesniaux V, Fullard L, Arendse H, Davison G, Markgraaff N, Auer R, Ehrhart F, Kraus G, Schuurman HJ (1999) A novel immunosuppressant, FTY720, induces peripheral lymphodepletion of both T- and B cells and immunosuppression in baboons. Transpl Immunol 7:149–157

Reichardt P, Dornbach B, Song R, Beissert S, Gueler F, Loser K, Gunzer M (2007a) Naive B cells generate regulatory T cells in the presence of a mature immunological synapse. Blood (in press)

Reichardt P, Gunzer F, Gunzer M (2007b) Analyzing the physicodynamics of immune cells in a 3-D collagen matrix. Methods Mol Biol (in press)

Reichardt P, Gunzer M (2006) The biophysics of T lymphocyte activation in vitro and in vivo. In Cell Cell Communication in the Nervous and Immune System, B Schraven, E Gundelfinger, C Seidenbecher, eds. (Berlin, Heidelberg, New York: Springer-Verlag Berlin), pp 199–218

Reif K, Ekland EH, Ohl L, Nakano H, Lipp M, Förster R, Cyster JG (2002) Balanced responsiveness to chemoattractants from adjacent zones determines B cell position. Nature 416:94–99

Reinhardt RL, Khoruts A, Merica R, Zell T, Jenkins MK (2001) Visualizing the generation of memory CD4 T cells in the whole body. Nature 410:101–105

Reis e Sousa C (2006) Dendritic cells in a mature age. Nat Rev Immunol 6:476–483

Richter A, Lohning M, Radbruch A (1999) Instruction for cytokine expression in T helper lymphocytes in relation to proliferation and cell cycle progression. J Exp Med 190:1439–1450

Rodriguez-Pinto D, Moreno J (2005) B cells can prime naive CD4(+) T cells in vivo in the absence of other professional antigen-presenting cells in a CD154-CD40-dependent manner. Eur J Immunol 35:1097–1105

Schiller M, Metze D, Luger TA, Grabbe S, Gunzer M (2006) Immune response modifiers—mode of action. Exp Dermatol 15:331–341

Schwab SR, Pereira JP, Matloubian M, Xu Y, Huang Y, Cyster JG (2005) Lymphocyte sequestration through S1P lyase inhibition and disruption of S1P gradients. Science 309:1735–1739

Schwickert TA, Lindquist RL, Shakhar G, Livshits G, Skokos D, Kosco-Vilbois MH, Dustin ML, Nussenzweig MC (2007) In vivo imaging of germinal centres reveals a dynamic open structure. Nature 446:83–87

Segal AW (2005) How neutrophils kill microbes. Annu Rev Immunol 23:197–223

Shiow LR, Rosen DB, Brdickova N, Xu Y, An J, Lanier LL, Cyster JG, Matloubian M (2006) CD69 acts downstream of interferon-alpha/beta to inhibit S1P(1) and lymphocyte egress from lymphoid organs. Nature 440:540–544

Shpacovitch VM, Varga G, Strey A, Gunzer M, Mooren F, Buddenkotte J, Vergnolle N, Sommerhoff CP, Grabbe S, Gerke V, Homey B, Hollenberg M, Luger TA, Steinhoff M (2004) Agonists of proteinase-activated receptor-2 modulate human neutrophil cytokine secretion, expression of cell adhesion molecules, and migration within 3-D collagen lattices. J Leukoc Biol 76:1–11

Snyderman R, Pike MC (1984) Chemoattractant receptors on phagocytic cells. Annu Rev Immunol 2:257–281

Sozzani S, Sallusto F, Luini W, Zhou D, Piemonti L, Allavena P, van Damme J, Valitutti S, Lanzavecchia A, Mantovani A (1995) Migration of dendritic cells in response to formyl peptides, C5a, and a distinct set of chemokines. J Immunol 155:3292–3295

Springer TA (1994) Traffic signals for lymphocyte recirculation and leukocyte emigration: the multistep paradigm. Cell 76:301–314

Steinman RM (1991) The dendritic cell system and its role in immunogenicity. Annu Rev Immunol 9:271–296

Steinman RM, Gutchinov B, Witmer MD, Nussenzweig MC (1983) Dendritic cells are the principal stimulators of the primary mixed leukocyte reaction in mice. J Exp Med 157:613

Steinman RM, Witmer MD (1978) Lymphoid dendritic cells are potent stimulators of the primary mixed leukocyte reaction in mice. Proc Natl Acad Sci USA 75:5132–5136

Stoll S, Delon J, Brotz TM, Germain RN (2002) Dynamic imaging of T cell-dendritic cell interactions in lymph nodes. Science 296:1873–1876

Suzuki N, Ohneda O, Minegishi N, Nishikawa M, Ohta T, Takahashi S, Engel JD, Yamamoto M (2006) Combinatorial Gata2 and Sca1 expression defines hematopoietic stem cells in the bone marrow niche. Proc Natl Acad Sci USA 103:2202–2207

Tadokoro CE, Shakhar G, Shen S, Ding Y, Lino AC, Maraver A, Lafaille JJ, Dustin ML (2006) Regulatory T cells inhibit stable contacts between CD4+ T cells and dendritic cells in vivo. J Exp Med 203:505–511

Takeda K, Kaisho T, Akira S (2003) Toll-Like Receptors. Annu Rev Immunol 21:335–376

Tang Q, Adams JY, Tooley AJ, Bi M, Fife BT, Serra P, Santamaria P, Locksley RM, Krummel MF, Bluestone JA (2006) Visualizing regulatory T cell control of autoimmune responses in nonobese diabetic mice. Nat Immunol 7:83–92

Tarlinton D (1998) Germinal centers: form and function. Curr Opin Immunol 10:245–251

Tauber AI (2003) Metchnikoff and the phagocytosis theory. Nat Rev Mol Cell Biol 4:897–901

ThurnerB, Haendle I, Röder C, Dieckmann D, Keikavoussi P, Jonuleit H, Bender A, Maczek C, Schreiner D, von Den Driesch P, Bröcker E-B, Steinman RM, Enk A, Kämpgen E, Schuler G (1999) Vaccination with Mage-3A1 peptide-pulsed mature, monocyte-derived dendritic cells expands specific cytotoxic T cells and induces regression of some metastases in advanced stage IV melanoma. J Exp Med 190:1669–1678

Tortora GJ, Grabowski SR (2000). The respiratory system. In Principles of anatomy and physiology, GJ Tortora, SR Grabowski, eds. (New York; Chichester; Weinheim; Brisbane; Singapore; Toronto: John Wiley & Sons, Inc.), pp 775–817

Trombetta ES, Mellman I (2005) Cell biology of antigen processing in vitro and in vivo. Annu Rev Immunol 23:975–1028

Tumpey TM, Lu X, Morken T, Zaki SR, Katz JM (2000) Depletion of lymphocytes and diminished cytokine production in mice infected with a highly virulent influenza A (H5N1) virus isolated from humans. J Virol 74:6105–6116

Underhill DM, Bassetti M, Rudensky A, Aderem A (1999) Dynamic interactions of macrophages with T cells during antigen presentation. J Exp Med 190:1909–1914

van der Merwe A, Davis SJ, Shaw AS, Dustin ML (2000) Cytoskeletal polarization and redistribution of cell-surface molecules during T cell antigen recognition. Semin Immunol 12:5–21

Verploegen S, Ulfman L, van Deutekom H, van Aalst C, Honing H, Lammers JW, Koenderman L, Coffer PJ (2005) Characterization of the role of CamKI Like Kinase (CKLiK) in human granulocyte function. Blood 106:1076–1083

von Andrian UH, M'Rini C (1998) In situ analysis of lymphocyte migration to lymph nodes. Cell Adhes Commun 6:85–96

von Andrian UH, Mempel TR (2003) Homing and cellular traffic in lymph nodes. Nat Rev Immunol 3:867–878

von Boehmer H, Aifantis I, Gounari F, Azogui O, Haughn L, Apostolou I, Jaeckel E, Grassi F, Klein L (2003) Thymic selection revisited: how essential is it? Immunol Rev 191:62–78

Warnock RA, Askari S, Butcher EC, von Andrian UH (1998) Molecular mechanisms of lymphocyte homing to peripheral lymph nodes. J Exp Med 187:205–216

Weninger W, Crowley MA, Manjunath N, von Andrian UH (2001) Migratory Properties of Naive, Effector, and Memory $CD8^+$ T Cells. J Exp Med 194:953–966

Wilson A, Trumpp A (2006) Bone-marrow haematopoietic stem-cell niches. Nat Rev Immunol 6:93–106

Witt C, Raychaudhuri S, Chakraborty AK (2005a) Movies, measurement, and modeling: the three Ms of mechanistic immunology. J Exp Med 201:501–504

Witt CM, Raychaudhuri S, Schaefer B, Chakraborty AK, Robey EA (2005b) Directed migration of positively selected thymocytes visualized in real time. PLoS Biol 3:e160

Wolkers MC, Brouwenstijn N, Bakker AH, Toebes M, Schumacher TN (2004) Antigen bias in T cell cross-priming. Science 304:1314–1317

Yokosuka T, Sakata-Sogawa K, Kobayashi W, Hiroshima M, Hashimoto-Tane A, Tokunaga M, Dustin ML, Saito T (2005) Newly generated T cell receptor microclusters initiate and sustain T cell activation by recruitment of Zap70 and SLP-76. Nat Immunol 6:1253–1262

Zinkernagel RM (1996) Immunology taught by viruses. Science 271:173–178

Zinkernagel RM, Doherty PC (1974) Restriction of in vitro T cell-mediated cytotoxicity in lymphocytic choriomeningitis within a syngeneic or semiallogeneic system. Nature 248:701–702

Ernst Schering Foundation Symposium Proceedings, Vol. 3, pp. 139–149
DOI 10.1007/2789_2007_073

Published Online: 29 February 2008

The Role of Diacylglycerol Kinases in T Cell Anergy

X.P. Zhong, B.A. Olenchock, G.A. Koretzky(✉)

Department of Pathology and Laboratory Medicine, Division of Rheumatology, Abramson Family Cancer Research Institute, University of Pennsylvania, 415 BRBII/III, 421 Curie Blvd., 19104 Philadelphia, USA
email: *Koretzky@mail.med.upenn.edu*

Abstract. Engagement of the T cell antigen receptor (TCR) results in the activation of multiple biochemical second messenger cascades that must be integrated for the appropriate T cell response. Once the critical TCR-stimulated signaling pathway is initiated by activation of protein tyrosine kinases, a series of adapter proteins is recruited that brings tyrosine-phosphorylated phospholipase Cγ1 into the vicinity of its substrate, phosphatidylinositol-4,5-bisphosphate, resulting in the formation of two second messengers, inositol-1,4,5-trisphosphate (IP_3) and diacylglycerol (DAG). Previous work from multiple laboratories has shown that the balance between signals downstream of IP_3 versus those downstream of DAG has profound effects on the fate of the stimulated T cells. In this report we summarize our recent data indicating that one key determinant of this balance of signals is the activity of members of the diacylglycerol kinase family, enzymes that convert DAG into phosphatidic acid.

1 Introduction

Naïve T cells become potent immune effectors after encountering specific antigen in the correct context. This conversion requires stimulation of both (1) the T cell antigen receptor (TCR) by peptide antigen presented by major histocompatibility complex (MHC) antigens and (2) the costimulatory receptors (e.g., CD28) by ligands present on antigen-presenting cells (APCs). Engagement of the TCR without concomitant stimulation of a coreceptor results not in development of T cell effector function but instead in a state of unresponsiveness known as anergy (Schwartz 2003). While this dichotomy in T cell responses has been long appreciated, the molecular determinants dictating whether a T cell becomes either an effector cell or anergic are only beginning to be understood.

Considerable insight into the biochemical basis of T cell anergy came from studies demonstrating that the balance of TCR-stimulated second messengers produced during T cell activation may play a critical role in whether the cell becomes an effector cell or is rendered unresponsive (Macian et al. 2004; Schwartz 2003). Some studies focused on the phos-

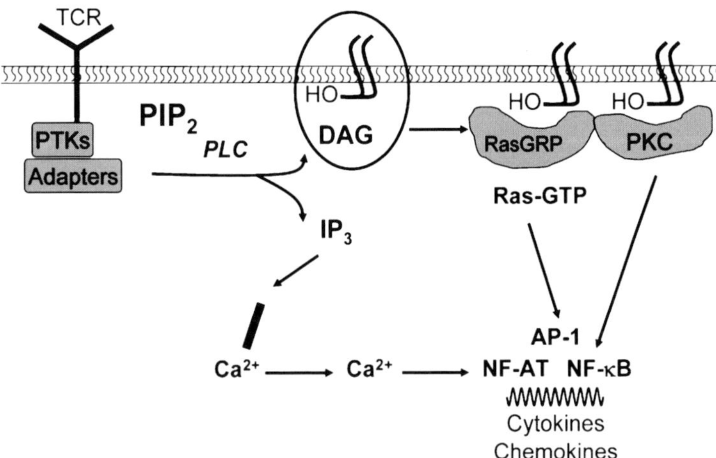

Fig. 1. TCR signaling

phatidylinositol second messenger cascade, a well-described signaling pathway initiated by engagement of the TCR (Jordan et al. 2003). The most proximal known signaling event that occurs when the TCR is engaged is activation of src (lck and/or fyn) family protein tyrosine kinases (PTKs). These enzymes phosphorylate the immunoreceptor tyrosine-based activation motifs (ITAMs) of the signaling components of the TCR. The phospho-ITAMs are then able to recruit ZAP-70, a syk family PTK, to the receptor. Activated ZAP-70 phosphorylates a number of key substrates, including several adapter proteins as well as the enzyme phospholipase Cγ1 (PLCγ1). The phosphorylated adapter proteins create and appropriately position a large multimolecular complex that brings active PLCγ1 into proximity with its substrate [the mem-

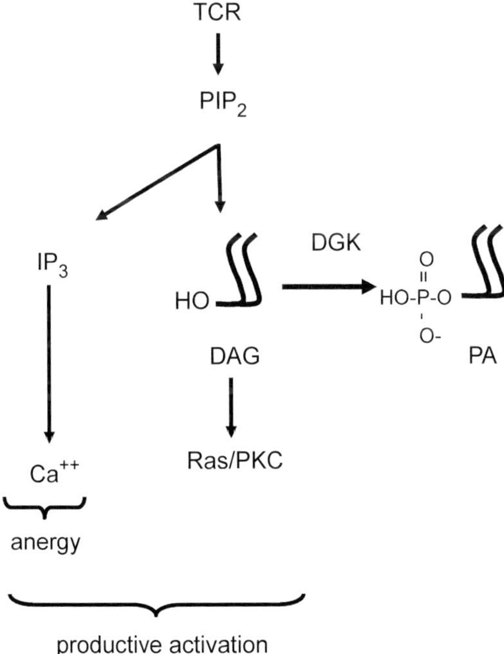

Fig. 2. Increased cytosolic T cell calcium without concomitant Ras/MAPK activation results in anergy

brane phospholipid phosphatidylinositol-4,5-bisphosphate (PIP_2)], allowing PLCγ1 to act on PIP_2, resulting in the formation of two second messengers: the soluble sugar inositol-1,4,5-trisphosphate (IP_3) and the lipid diacylglycerol (DAG). IP_3 binds to its receptor on the endoplasmic reticulum, inducing the release of calcium from intracellular stores. This calcium then initiates a series of additional second messenger events. DAG is also critical for T cell responses as it binds to and activates various proteins, including members of the protein kinase C (PKC) family and RasGrp1. RasGrp1 is an exchange factor that enhances the release of GDP from Ras, thus allowing Ras to bind GTP and become activated. Active Ras then initiates a signaling cascade, stimulating members of the mitogen-activated protein kinase (MAPK) family (Dower et al. 2000; Ebinu et al. 2000; Fig. 1).

A number of experimental systems have now shown that T cell effector function requires that calcium and MAPK signals be balanced. In particular, increased cytosolic T cell calcium without concomitant Ras/MAPK activation fails to result in the development of T cell effector function but results instead in long-lived T cell anergy (Jenkins et al. 1987; Macian et al. 2002; Fig. 2). This finding led us to hypothesize that one means by which T cells could potentially discriminate between sig-

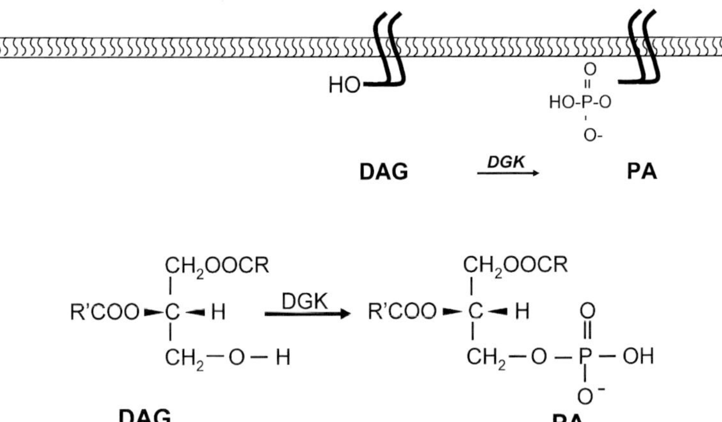

Fig. 3. DGK converts DAG to PA

nals that should elicit an anergic response versus development of T cell effector function would be control over DAG, the upstream regulator of the Ras/MAPK pathway (Kazanietz 2005). We reasoned further that a possible method for this control would be regulation of diacylglycerol kinases (DGKs), the enzyme family that terminates DAG signaling by converting DAG into phosphatidic acid (PA) (Luo et al. 2004; van Blitterswijk and Houssa 2000; Fig. 3).

2 Gain- and Loss-of-Function Studies Examining DGKζ

There are ten members of the DGK family found in mammals (Fig. 4). Each family member possesses a kinase domain as well as multiple C1 (DAG binding) domains. DGKs differ considerably in their structures (with various protein interaction and subcellular localization domains) and in their tissue distribution, with α and ζ being the predominant isoforms in T cells (Sanjuán et al. 2003; Zhong et al. 2002).

We began our studies using the Jurkat model system, a cell line that has faithfully recapitulated the most important observations regarding the molecular events associated with TCR signaling in normal primary T cells. We focused first on DGKζ by evaluating the impact on TCR

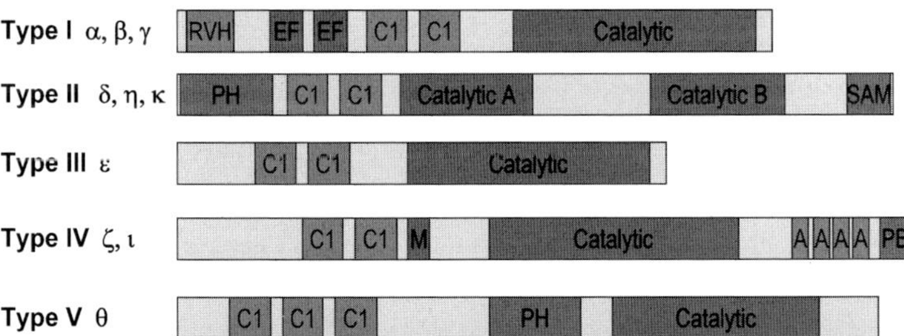

Fig. 4. DGK isoforms found in mammals. *RVH*, recoverin homology domain; *EF*, EF-hand; *C1*, cysteine-rich domain 1; *SAM*, sterile α motif; *PH*, pleckstrin homologous domain; *M*, MARCKS (myristoylated alanine-rich protein kinase C substrate) motif; *A*, Ankyrin-repeat; *PB*, PDZ (PSD-95, DLG, ZO-1 domain) binding motif

signaling in Jurkat cells when this DGK isoform was overexpressed by transfection of cDNA. Our experimental protocol was to express high levels of DGKζ in Jurkat cells and then compare the outcome of TCR engagement on these cells compared to cells transfected with control cDNA. As predicted, overexpression of DGKζ had no impact on the ability of the TCR to stimulate a calcium response, as DGK functions downstream of the production of this second messenger. In contrast, DGKζ overexpression had a profound effect on the TCR-stimulated Ras/MAPK pathway; signaling via this pathway was quickly terminated. A subsequent structure/function analysis revealed that this effect required both the DAG-binding and enzymatic region of DGKζ (Zhong et al. 2002).

To take this work beyond the Jurkat model system, we next generated a retrovirus encoding DGKζ cDNA along with a green fluorescent protein reporter. We used this virus to infect bone marrow precursors of wildtype mice and used the manipulated bone marrow to reconstitute lethally irradiated wildtype mice. Mice were sacrificed, and the thymi were analyzed for T cell development 8 weeks later. The only mature T cells present in the mice were those that were not transduced by the DGKζ virus. T cell development was arrested at the double-negative stage in cells arising from DGKζ-overexpressing precursors, likely due to failed Ras/MAPK signaling.

We complemented our gain-of-function studies with a loss-of-function approach. We generated mice in which the DGKζ gene was targeted by homologous recombination; these mice were viable and fertile, demonstrating no overt abnormal phenotype. Gross T cell development in the DGKζ$^{-/-}$ animals was also unaffected. However, ex vivo analysis of mature T cells from these mice revealed increased Ras/MAPK signaling following TCR engagement without an alteration in calcium signaling. This increased signaling was sufficient to result in enhanced proliferation and cytokine production in the DGKζ$^{-/-}$ cells. The physiologic significance of these ex vivo findings was shown in experiments where wildtype and DGKζ$^{-/-}$ mice were challenged with lymphocytic choriomeningitis virus (LCMV). We found that the genetically altered mice cleared virus more readily than their littermate controls and made a more potent antiviral T cell response. Preliminary experiments suggest that a model tumor is rejected more potently in the setting of DGKζ deficiency. Collectively, these experiments suggest that DGKζ may be a physiologically important negative regulator of T cell effector function (Zhong et al. 2003). This finding is particularly interesting given the fact that we did not manipulate DGKα (the other major isoform in T cells) expression in these experiments.

Additional preliminary evidence supports the possibility that DGKζ functions as a controller of T cell activation. As DGKζ$^{-/-}$ mice age, they have a propensity to develop autoantibodies. Although the animals do not manifest overt disease on a C57/Bl6 background, studies are underway to determine if DGKζ deficiency may synergize with other genetic abnormalities on an autoimmune disease-prone background. We have also found that when the DGKζ$^{-/-}$ mice are bred with mice expressing a fixed TCR transgene, both positive and negative selection are skewed, consistent with increased strength of TCR signaling. Thus, it is likely that when mice are not forced to use a specific TCR, DGKζ deficiency alters the TCR repertoire, perhaps selecting receptors with decreased affinity for self-MHC.

3 The Phosphatidylinositol Pathway and T Cell Anergy

One means to induce T cell anergy is by stimulation of T cells via their TCR alone, without costimulation via other receptors such as CD28 (Schwartz 2003). We wondered, therefore, if one consequence of costimulation might be the potentiation of DAG signaling, thus leading to enhanced Ras/MAPK activation. Unfortunately, it is difficult to directly assess TCR (and TCR plus costimulation)-induced changes in cellular DAG because of the high baseline DAG mass in T cells. We therefore chose to assess DAG signaling capacity indirectly by measuring consumption of PIP_2 (the DAG precursor) and production of IP_3 (a second messenger generated in equimolar concentration to DAG) and PA (the product of the action of DGK on DAG). For these experiments, we stimulated wildtype T cells via their TCRs plus CD28 by incubating freshly isolated lymph node cells with anti-TCR antibody along with APCs (expressing CD28 ligand). To deliver a signal via the TCR alone, we used the same stimulation conditions but also included CTLA4Ig, a reagent that effectively competes for CD28 ligands, thus blocking costimulation.

The results of these experiments were quite striking. We found that while stimulation of T cells via the TCR alone resulted in significant consumption of PIP_2, this event was markedly enhanced when both the TCR and CD28 were engaged. As expected, costimulation resulted in significantly increased IP_3 production. However, both the magnitude and kinetics of PA production did not differ in costimulated versus TCR-alone stimulated cells. These studies collectively suggest that one consequence of stimulating T cells via their TCR plus CD28 is enhanced DAG signaling (Olenchock et al. 2006).

4 DGKs and "Anergic Signaling"

Several years ago an important study profiled the transcriptionally active genes in T cells that had been induced to become anergic. Among the genes reported from this screen was the gene encoding DGKα, the other major DGK isoform in T cells (Macian et al. 2002). We confirmed and extended this observation by analyzing mRNA levels of DGKα and DGKζ in resting, effector, and anergic T cells. Whereas levels of these enzymes are relatively high in resting cells, mRNA for both isoforms falls precipitously in cells induced to have effector function and rises in cells induced to be anergic (Olenchock et al. 2006). We speculated, therefore, that overexpression of either DGKζ or DGKα would confer an "anergic" signaling phenotype. To test this idea, we transduced Jurkat cells with either DGK isoform and then analyzed the cells for calcium and Ras/MAPK signaling as well as events downstream of these pathways, such as activation of nuclear factor of activated T cells, AP-1, and NF-κB. These studies revealed that Jurkat cells, which are easily activated to exhibit features of effector cells, lose this capacity when either DGKα or DGKζ is expressed at a high level (Olenchock et al. 2006).

5 DGKs and Functional Anergy

Another series of experiments was designed to inquire more directly about whether manipulation of DGK levels would impact anergy in freshly isolated primary murine T cells. For these studies it was necessary to generate mice deficient in DGKα to complement the animals we had already generated that were DGKζ deficient. DGKα-deficient animals were generated using standard gene-targeting approaches. Many findings were similar to what we observed in the DGKζ$^{-/-}$ animals. In the DGKα$^{-/-}$ mice, gross T cell development appeared normal. Ex vivo experiments indicated a similar impact in assays of calcium and Ras/MAPK signaling, and DGKα$^{-/-}$ mice also exhibited an increase in TCR-induced cytokine production. These findings indicate that the two isoforms each contribute to function and that neither is entirely redundant (Olenchock et al. 2006).

The obvious next experiment was to generate primary peripheral T cells lacking both DGKζ and DGKα to determine the phenotype in the absence of both isoforms. This analysis, however, was impossible to perform, as mice doubly deficient in DGKα and DGKζ fail to generate sufficient numbers of T cells for study, and the few cells that are generated appear to be selected aberrantly in the thymus. However, there is a commercially available reagent (R59022) that has some selectivity as a DGKα inhibitor. We therefore chose to test the impact of DGK deficiency on functional anergy using a combination of genetic and pharmacologic approaches.

The experimental paradigm for these studies was to isolate primary peripheral T cells from wildtype mice, DGKζ$^{-/-}$ mice, and DGKα$^{-/-}$ mice. These cells were incubated with anti-TCR plus APCs in the absence or presence of CTLA4Ig (to block costimulation and hence induce anergy) and with R59022 to inhibit DGKα in the wildtype or DGKζ-deficient cells. The DGKα inhibitor phenocopied quite closely the DGKα-deficient cells and had an additive effect on cells lacking DGKζ, suggesting that its action is relatively specific for the α-isoform (Olenchock et al. 2006). For each stimulation condition we loaded the responding T cells with CFSE (carboxy fluorescein diacetate succinimidyl ester), a dye that binds irreversibly to cellular proteins and that can be measured by flow cytometry. Cellular proliferation can be detected by CFSE dye dilution, as cellular division results in halving of the amount of CFSE remaining in each daughter cell.

The results of these experiments revealed that loss of either DGKζ or DGKα alone had a considerable impact on cellular proliferation under conditions that should have induced anergy. Thus, while wildtype T cells failed to divide if stimulated via the TCR alone (with anti-TCR antibodies plus APCs with CTLA4Ig), both DGKζ- and DGKα-deficient cells underwent several rounds of proliferation under these same stimulation conditions. The response of DGKζ-deficient cells treated with the DGKα pharmacologic inhibitor were even more striking, as TCR stimulation in the setting of CD28 blockade was just as effective in initiating proliferation as was stimulation of wildtype T cells with both TCR and CD28 agonists. These studies suggest, at least when assessed ex vivo, that inhibition of both DGK isoforms effectively converts an anergic stimulus into a response ordinarily seen only following TCR plus CD28 engagement (Olenchock et al. 2006).

Our final series of experiments tested the impact of DGK deficiency on anergy induction in vivo. For this study we made use of the DGKα-deficient animals, comparing them to wildtype littermate controls. Animals were injected with a superantigen, staphylococcal enterotoxin B (SEB), which has been shown previously to induce rapid clonal expansion of T cells bearing Vβ8 TCRs. A contraction phase follows, during which most of the responding cells undergo apoptosis. The surviving cells that had responded initially to the superantigen challenge become unresponsive to restimulation with SEB in vitro. We found, as expected, that the wildtype mice exhibit anergy under these conditions, manifested both by a markedly diminished proliferative response and by failure to secrete interleukin 2. In contrast, cells recovered from the DGKα-deficient mice responded briskly to antigenic rechallenge in vitro, complementing our findings that demonstrated failed anergy induction by blocking CD28 signaling (Olenchock et al. 2006).

6 Conclusion

Our studies reveal that both DGKα and DGKζ are important negative regulators of TCR signaling and T cell activation by terminating DAG-mediated signaling. DGKα and DGKζ activities contribute to T cell anergy in vitro and in vivo and play an important role for self-tolerance.

References

Dower NA, Stang SL, Bottorff DA, Ebinu JO, Dickie P, Ostergaard HL, Stone JC (2000) RasGRP is essential for mouse thymocyte differentiation and TCR signaling. Nat Immunol 1:317–321

Ebinu JO, Stang SL, Teixeira C, Bottorff DA, Hooton J, Blumberg PM, Barry M, Bleakley RC, Ostergaard HL, Stone JC (2000) RasGRP links T-cell receptor signaling to Ras. Blood 95:3199–3203

Jenkins MK, Pardoll DM, Mizuguchi J, Chused TM, Schwartz RH (1987) Molecular events in the induction of a nonresponsive state in interleukin 2-producing helper T-lymphocyte clones. Proc Natl Acad Sci USA 84:5409–5413

Jordan MS, Singer AL, Koretzky GA (2003) Adaptors as central mediators of signal transduction in immune cells. Nat Immunol 4:110–116

Kazanietz MG (2005) Targeting protein kinase C and "non-kinase" phorbol ester receptors: emerging concepts and therapeutic implications. Biochim Biophys Acta 1754:296–304

Luo B, Regier DS, Prescott SM, Topham MK (2004) Diacylglycerol kinases. Cell Signal 16:983–989

Macian F, Garcia-Cozar F, Im SH, Horton HF, Byrne MC, Rao A (2002) Transcriptional mechanisms underlying lymphocyte tolerance. Cell 109:719–731

Macian F, Im SH, Garcia-Cozar FJ, Rao A (2004) T-cell anergy. Curr Opin Immunol 16:209–216

Olenchock BA, Guo R, Carpenter JH, Jordan M, Topham MK, Koretzky GA, Zhong XP (2006) Disruption of diacylglycerol metabolism impairs the induction of T cell anergy. Nat Immunol 7:1174–1181

Sanjuán MA, Pradet-Balade B, Jones DR, Martínez-A C, Stone JC, Garcia-Sanz JA, Mérida I (2003) T cell activation in vivo targets diacylglycerol kinase α to the membrane: a novel mechanism for Ras attenuation. J Immunol 170:2877–2883

Schwartz RH (2003) T cell anergy. Annu Rev Immunol 21:305–334

van Blitterswijk WJ, Houssa B (2000) Properties and functions of diacylglycerol kinases. Cell Signal 12:595–605

Zhong XP, Hainey EA, Olenchock BA, Zhao H, Topham MK, Koretzky GA (2002) Regulation of T cell receptor-induced activation of the Ras-ERK pathway by diacylglycerol kinase zeta. J Biol Chem 277:31089–31098

Zhong XP, Hainey EA, Olenchock BA, Jordan MS, Maltzman JS, Nichols KE, Shen H, Koretzky GA (2003) Enhanced T cell responses due to diacylglycerol kinase zeta deficiency. Nat Immunol 4:882–890

Ernst Schering Foundation Symposium Proceedings, Vol. 3, pp. 151–167
DOI 10.1007/2789_2007_063

Published Online: 18 December 2007

Molecular Mechanisms that Control Leukocyte Extravasation Through Endothelial Cell Contacts

D. Vestweber(✉)

Max Planck Institute of Molecular Biomedicine, Röntgenstr. 20, 48149 Münster, Germany

email: *vestweb@mpi-muenster.mpg.de*

Abstract. Leukocyte extravasation and entry into tissue forms the basis for inflammatory reactions and lymphocyte surveillance. After docking at the blood vessel wall at sites of exit leukocytes migrate through the endothelial cell layer and the underlying basement membrane, a process described as diapedesis. In recent years, several endothelial membrane proteins that which participate in this process have been identified. This review focuses on three membrane proteins located at endothelial cell contacts that are involved in the regulation of leukocyte diapedesis. The endothelial cell selective adhesion molecule (ESAM) at endothelial tight junctions and the vascular endothelial receptor-type protein tyrosine phosphatase (VE-PTP), a protein associating with VE-cadherin, both seem to control the integrity of endothelial cell contacts during diapedesis.

CD99 and the distantly related CD99L2 are leukocyte membrane proteins that do not belong to any known protein family. They are expressed at endothelial cell contacts and participate in the migration of leukocytes through endothelium and basement membrane.

1 Introduction

Migration of leukocytes into tissue is a key element of innate and adaptive immunity. While the docking of leukocytes to the blood vessel wall has been analysed intensively for more than 15 years and is now well understood, rather little is known about the mechanisms underlying the actual transmigration of leukocytes through the vessel wall (diapedesis). Even a basic question such as whether leukocytes migrate through openings between adjacent endothelial cells (junctional pathway) or through the body of an endothelial cell (transcellular pathway) is still a matter of intensive debate. Convincing evidence has been published that suggests that both pathways exist but quantification of in vitro studies clearly demonstrates that the majority of leukocytes migrate through endothelium at or at least close to endothelial cell contacts. Almost all endothelial cell surface proteins that are known to be involved in diapedesis are found at endothelial cell contacts. In most cases it is not known in detail how these proteins function: some may support directly the migration of leukocytes through the cleft between endothelial cells, some may act by triggering signals in leukocytes that indirectly support migration, some may trigger the opening of endothelial cell contacts, and others represent inter-endothelial cell adhesion molecules that form barriers for emigrating leukocytes, which are down-regulated during the extravasation process.

2 Presently Known Adhesion Receptors at Endothelial Cell Contacts Involved in Leukocyte Diapedesis

PECAM-1 was the first endothelial cell contact protein that was found to support leukocyte emigration. It is a member of the immunoglobulin supergene family (Ig-SF) containing six Ig-domains and it is not confined to any type of junctional structures. Besides endothelial cells, PECAM-1 is expressed on platelets, neutrophils, monocytes and particular T cell subsets. Antibodies against PECAM-1 as well as a PECAM-Fc fusion protein could block transendothelial migration of neutrophils in vitro and neutrophil extravasation in vivo in mice and rats (Muller et al. 1993; Liao et al. 1997). Although a first report on C57Bl/6 mice, deficient for the PECAM-1 gene, revealed no significant reduction in neutrophil extravasation (Duncan et al. 1999), the same gene deficiency caused a clear reduction in neutrophil emigration when the mice were backcrossed into the FVB/n strain (Schenkel et al. 2004). The mechanism by which PECAM-1 participates in leukocyte extravasation is not yet known in all detail. It has been reported that the migration of monocytes through cell contacts between adjacent endothelial cells require different Ig-domains of PECAM-1 than the migration through the basement membrane (Liao et al. 1995), suggesting PECAM-1 could act at two different steps of diapedesis. However, only the 'second' step has been confirmed in vivo. Neutrophils were shown in vivo to get stuck between endothelial cells and basement membrane when PECAM-1 was blocked with a monoclonal antibody (Wakelin et al. 1996). Increased expression of the integrin $\alpha_6\beta_1$ on the surface of neutrophils was found to be involved in this process (Dangerfield et al. 2002). Indeed, others showed that endothelial PECAM-1 is rather passively involved as a homophilic ligand for neutrophil PECAM-1, which in turn is needed as signal transducer and stimulus for leukocyte integrin activation (O'Brien et al. 2003).

Human monocytes were found to use CD99, a glycoprotein expressed on leukocytes as well as at endothelial cell contacts, to migrate in vitro through the layer of endothelial cells. CD99 is a protein with no structural similarity to any known protein family. CD99 was shown to act sequentially with PECAM-1 at a step that is required later than the 'first' step mediated by PECAM-1 (Schenkel et al. 2002). This study suggested for the first time two consecutive molecular steps during diapedesis. Extending these observations, mouse CD99 was recently shown to be relevant for the extravasation of lymphocytes in vivo (Bixel et al. 2004). CD99 and a distantly related second membrane protein, CD99L2, will be discussed later in this review.

JAM-A, another member of the Ig-SF with only two Ig domains, was the first tight junction-associated endothelial protein found to be involved in monocyte and neutrophil extravasation, based on the inhibitory effect of a monoclonal antibody against JAM-A (Martin-Padura et al. 1998; Del Maschio et al. 1999). JAM-A is not specific for endothelial cells, it is also found on epithelial cells and in vitro studies documented that polyclonal antibodies against JAM-A can interfere with the formation of epithelial cell contacts (Liu 2000). Surprisingly, the same antibodies did not inhibit the migration of leukocytes through endothelial or epithelial cell monolayers (Liu et al. 2000). It is still unclear how JAM-A participates in leukocyte extravasation. Various studies with JAM-A gene-deficient mice come to very different conclusions. Two of these even suggest that it is JAM-A on myeloid leukocytes that is involved in the diapedesis step, whereas JAM-A on endothelial cells was found to be dispensable, as demonstrated with mice selectively deficient for JAM-A on endothelial cells (Cera et al. 004; Corada et al. 2005). It is the motility of leukocytes that seems to be affected by the lack of JAM-A. Unexpectedly, dendritic cells lacking JAM-A migrate faster through lymphatic endothelial cell layers in vitro and migrate more efficiently into lymph node tissue in vivo (Cera et al. 2004), whereas neutrophils lacking JAM-A are slowed down in extravasation. In contrast to these two studies, participation of endothelial JAM-A in neutrophil extravasation was clearly demonstrated in a liver model for ischemia-reperfusion injury (Khandoga et al. 2005). More recently it was found that JAM-A and PECAM-1 indeed act in a sequential manner. As shown by confocal microscopy of whole mounts of cremaster tissue, neutrophils became trapped between endothelial cells in JAM-A$^{-/-}$ mice, whereas neutrophil accumulation in PECAM-1$^{-/-}$ mice was observed between endothelium and basement membrane (Woodfin et al. 2007).

The two closely related proteins JAM-B and JAM-C are also expressed at endothelial cell contacts (Aurrand-Lions et al. 2001) and a soluble JAM-C-Fc fusion protein inhibited neutrophil extravasation in vivo (Chavakis et al. 2004a) and neutrophil migration through epithelial cell layers (Zen et al. 2004). Interestingly, each of these three JAM proteins seems to be able to interact with leukocyte integrins (Chavakis et al. 2004b). In addition they bind in a homophilic fashion and JAM-B binds very efficiently to JAM-C (Arrate et al. 2001), providing numerous ways by which these molecules could participate in leukocyte extravasation. Another protein distantly related to the JAM proteins and also involved in leukocyte extravasation, is endothelial cell selective adhesion molecule (ESAM) that will be discussed later in this review.

A third type of Ig-SF adhesion protein at endothelial cell contacts is ICAM-2, a long known endothelial and leukocyte antigen, that is a ligand of the leukocyte integrin LFA-1. Based on in vitro studies ICAM-2 had been suggested to participate in the transendothelial migration of lymphocytes (Reiss et al. 1998). Recently, a monoclonal antibody (mAb) against ICAM-2 had been shown to block recruitment of neutrophils into inflamed peritoneum as well as extravasation of neutrophils in the interleukin (IL)-1β stimulated cremaster muscle (Huang et al. 2006). Since the cremaster experiments were analysed by intravital microscopy, it could be excluded that ICAM-2 would be involved in leukocyte capturing to the endothelium, indicating that it is the diapedesis step where ICAM-2 is necessary. The same study showed that a mAb against ICAM-2 strongly inhibited neutrophil recruitment into inflamed peritoneum beyond the inhibitory effect found in PECAM-1 deficient mice suggesting that the two proteins act at different steps in the diapedesis process.

The nectins represent a group of junctional adhesion molecules that are again members of the Ig-SF family, but in contrast to JAM these are located at adherens junctions. One of their members, the poliovirus receptor (PVR), is found at endothelial contacts, binds to the leukocyte antigen DNAM-1 (CD226) and participates in the migration of monocytes through endothelial cell layers in culture (Reymond et al. 2004).

In summary, with PECAM-1, CD99, the JAM proteins, ICAM-2 and PVR, we know five different types of endothelial cell contact proteins, that all participate in leukocyte diapedesis. As for PECAM-1, CD99 and ICAM-2 we have first indications that some of these proteins might act at various steps in a cascade of molecular interactions that mediate the diapedesis process. Central questions are: (1) Which of these proteins or what other proteins control and or mediate the opening of the endothelial cell contacts? (2) Which proteins serve as support for the migration of leukocytes through endothelial cell contacts? All of these five types of cell contact proteins are able to mediate homophilic molecular interactions, although some of these interactions are rather weak and not sufficient to support cell aggregation of transfected cells (e.g., JAM-A). Whether any of them is involved in the control of the opening of endothelial cell contacts is unclear.

In contrast to the endothelial cell contact proteins described above, VE-cadherin does not support leukocyte extravasation. VE-cadherin forms a barrier for leukocytes, stabilizing endothelial cell contact stability. Leukocytes that want to cross this barrier need to down-regulate the function of VE-cadherin. It is not yet known in detail how this is achieved.

3 VE-PTP, a VE-Cadherin-Associated Receptor Type Tyrosine Phosphatase that Regulates VE-Cadherin Function

Several cell surface proteins at endothelial cell contacts are able to mediate homophilic adhesion, however, VE-cadherin is the only cell adhesion molecule known to be essential for the maintenance of inter-endothelial cell contacts. In analogy to the importance of E-cadherin for the integrity of epithelial cell layers, it was found that antibodies against human VE-cadherin dissociate the contacts of endothelial cells in culture (Lampugnani et al. 1992). VE-cadherin is expressed at adherens junctions and associates, like most other cadherins, with the cytoplasmic catenins. β-catenin and placoglobin (γ-catenin) connect VE-cadherin to α-catenin, an important connection for the stability of cadherin-mediated adhesions. It is thought that this complex is needed for the interaction with the cytoskeleton, although a direct interaction with the actin cytoskeleton has recently been questioned (Yamada et al. 2005).

VE-cadherin-mediated endothelial cell contacts represent a barrier for extravasating leukocytes in vivo, since a mAb against VE-cadherin administered intravenously in mice accelerated neutrophil migration into the inflamed peritoneum (Gotsch et al. 1997). The importance of VE-cadherin for the integrity of the blood vessel endothelium in vivo was confirmed with another antibody that induced a concentration- and time-dependent increase in vascular permeability in heart and lungs (Corada et al. 1999). Real-time imaging of a VE-cadherin-GFP fusion protein in cultured human umbilical vein endothelial cells (HUVEC) revealed that this protein is removed from cell contact sites during leukocyte transmigration and re-appears afterwards, possibly via diffusion in the plane of the membrane (Shaw et al. 2001).

Changes in tyrosine phosphorylation of cadherin/catenin complexes with modulation of cadherin-mediated adhesion have been reported in several publications (Daniel and Reynolds 1997). Such correlations were also found for VE-cadherin. Increased confluence of endothelial cells in culture was accompanied by a decrease in VE-cadherin/catenin tyrosine phosphorylation and a decrease of staining of adherens junctions by anti-phosphotyrosine antibodies (Lampugnani et al. 1997). Thrombin stimulation of human endothelial cells promotes the dissociation of the phosphatase SHP2 from β-catenin, which correlates with an increase in catenin phosphorylation (Ukropec et al. 2000). Furthermore, stimulation of HUVEC with vascular endothelial growth factor (VEGF) induced tyrosine phosphorylation of VE-cadherin and the associated catenins (except α-catenin) and increased endothelial permeability (Esser et al. 1998).

Recently, we found an endothelial specific receptor-protein tyrosine phosphatase (R-PTP), called VE-PTP, to associate specifically and selectively with VE-cadherin (Nawroth et al. 2002). VE-PTP had originally been identified as an endothelial specific vascular endothelial (VE)-PTP (Fachinger et al. 1999) that is highly homologous to human HPTP-β and that associates with the tyrosine kinase receptor Tie-2. The association of VE-PTP with VE-cadherin is fundamentally different from other R-PTP/cadherin associations as it is not mediated via a catenin and or via cytoplasmic cadherin domains, but instead via extracellular domains of VE-cadherin and VE-PTP. Expression of VE-PTP in triple-transfected CHO cells reversed the tyrosine phosphorylation of VE-cadherin elicited by vascular endothelial growth factor-receptor 2 (VEGFR-2). Expression of VE-PTP under an inducible promoter in CHO cells transfected with VE-cadherin and VEGFR-2, increased VE-cadherin-mediated barrier integrity of a cellular monolayer. Thus, VE-PTP is a transmembrane binding partner of VE-cadherin that associates through an extracellular domain, reverses VEGFR-2 mediated phosphorylation of VE-cadherin and influences cell layer permeability. The highly specific interaction of VE-PTP with VE-cadherin further supports the hypothesis that tyrosine phosphorylation of components of the VE-cadherin/catenin complex or of factors in its vicinity is involved in the regulation of VE-cadherin-mediated cell contacts.

Indeed, we found recently that knocking down VE-PTP expression in endothelial cells leads to increased permeability across the endothelial cell layer as well as to increased transendothelial migration of neutrophils (L.M. Nottebaum et al., unpublished results). Thus VE-cadherin is indeed required for the adhesive function of VE-cadherin and the maintenance of endothelial cell contacts. Furthermore we could show that the migration of leukocytes through endothelial cell layers increased if VE-PTP expression was inhibited. Docking of leukocytes to endothelial cells leads to the dissociation of VE-cadherin from VE-PTP strongly suggesting that VE-PTP is indeed involved in the process of leukocyte diapedesis (L.M. Nottebaum et al., unpublished results).

Due to the binding of VE-PTP to Tie-2 and VE-cadherin, both essential genes in angiogenesis, we analysed whether VE-PTP would have an important role in embryonic development. To this end we analysed mice with a disrupted VE-PTP gene, that gave rise to a secreted soluble form of VE-PTP, lacking any enzyme activity, which did not stay associated with the surface of the cells. Heterozygous mice were normal, but homozygous mutant mice showed severe vascular malformations causing embryonic lethality at 10 days of gestation. Although blood vessels were initially formed, the intraembryonic vascular system soon deteriorated. Blood vessels in the yolk sac developed into dramatically enlarged cavities. In explant cultures of mutant allantoides, endothelial cells were found next to vessel structures, growing as cell layers. It is conceivable that the lack in vessel stability might be due to defects in the activity of VE-cadherin, although other binding partners or substrates of VE-PTP cannot be excluded. We observed no signs for enhanced endothelial apoptosis or proliferation in blood vessels of VE-PTP mutant embryos. We conclude that the activity of VE-PTP is not required for the initial formation of blood vessels, yet it is essential for their maintenance and remodelling (Baumer et al. 2006).

4 ESAM Controls Endothelial Cell Contacts During Leukocyte Extravasation

ESAM has been identified by two laboratories using two different and independent experimental approaches. A differential cDNA hybridization approach defined ESAM as an endothelial gene possibly up-regulated during tube formation of HUVEC in culture (Hirata et al. 2001). Parallel to these studies, ESAM was identified in our group as an endothelial junction protein, by screening mAb for staining of endothelial cell contacts. Purifying and sequencing of the protein formed the basis for cloning the antigen (Nasdala et al. 2002). We showed that ESAM is specifically expressed by endothelial cells and platelets, but not by leukocytes. Analysing the subcellular distribution by immunogold electron microscopy revealed that ESAM is strictly associated with endothelial tight junctions.

ESAM is structurally related to the JAM family, although the various JAM proteins are clearly more closely related to each other than to ESAM. The JAM proteins bind to different PDZ domain scaffolding proteins than does ESAM, indicating that both types of proteins interact with different signal mediators and probably function in different ways (Ebnet et al. 2003; Wegmann et al. 2004).

Analysing gene-deficient mice it was found, that ESAM is involved in tumour angiogenesis (Ishida 2003) although no obvious defects during embryonal angiogenesis have been observed. We analysed a possible function of ESAM in leukocyte extravasation. Despite employing various mAb as well as polyclonal antibodies against ESAM in different inflammation models, it was not possible to block the migration of activated T cells into inflamed skin or of neutrophils into inflamed peritoneum. However, ESAM-deficient mice had a clear delay in neutrophil extravasation at early time points (2 h) after stimulation with either thioglycollate or cytokines such as TNF-α or IL-1β. Leukocyte extravasation was delayed at the diapedesis step, as shown by intravital microscopy of cytokine-stimulated cremaster tissue (Wegmann et al. 2006).

For the other endothelial adhesion receptors at endothelial cell contacts, the detailed mechanism by which ESAM participates in leukocyte extravasation is not yet known. However some aspects of this mechanism could be clarified. The first concerned a possible involvement of ESAM on platelets. As platelets represent the only other cell type besides endothelial cells that express ESAM, and because platelets are known to contribute to neutrophil extravasation, we tested whether the depletion of platelets would still affect neutrophil recruitment into inflamed peritoneum in ESAM deficient mice. Indeed, we found a clear reduction of neutrophil extravasation in wild-type mice but this reduction was the same when we analysed ESAM-deficient mice. Thus, platelets contribute to neutrophil recruitment independently of whether they express ESAM or not (Wegmann et al. 2006). It follows that only ESAM at endothelial tight junctions is involved in neutrophil diapedesis.

Another obvious question we asked was whether the lack of ESAM would affect vascular permeability. We found that the steady-state permeability for the plasma protein-adsorbed dye Evans blue was unchanged. However, the VEGF-triggered increase of vascular permeability in the skin was dramatically delayed in ESAM$^{-/-}$ mice, suggesting that ESAM is involved in the signalling pathway that connects VEGF signalling to the opening of endothelial junctions (Wegmann et al. 2006). As a first hint towards further signalling steps we found that knocking down the expression of ESAM in cultured endothelial cells leads to reduced levels of the GTPase Rho.

The permeability experiments suggest that ESAM may be generally involved in mechanisms that control and mediate the opening of endothelial contacts. In this context it is interesting that the lack of ESAM delayed extravasation of neutrophils into inflamed peritoneum only, and not the recruitment of activated T cells into inflamed skin (Wegmann et al. 2006). In order to exclude that this different result was due to tissue differences or differences in the timing of extravasation of the two different types of leukocytes, we recruited neutrophils and lymphocytes with a mixture of thioglycollate and the chemokine CCL19, both of which were injected into the peritoneum. In this way we recruited T cells and neutrophils together and at the same time into the same tissue. Even under these conditions we found that only the recruitment of neutrophils but not of lymphocytes was delayed. Thus, by whatever mechanism ESAM participates in the opening of endothelial cell contacts during neutrophil extravasation, this mechanism is not required for lymphocytes.

5 CD99 and CD99L2 Participate in Leukocyte Extravasation into Inflamed Tissue In Vivo

As outlined in the introduction CD99, a long known leukocyte cell surface antigen of unknown function had been discovered a few years ago to be expressed at endothelial cell contacts. A mAb against human CD99 was shown to inhibit monocyte diapedesis through the monolayer of HUVEC cells in vitro (Schenkel et al. 2002). Furthermore, this report presented evidence that anti-PECAM-1 antibodies trapped monocytes just above endothelial cell contacts, whereas anti-CD99 antibodies trapped leukocytes between the contacts of endothelial cells, suggesting a sequential action of PECAM-1 and CD99. This was recently confirmed in another in vitro study with human neutrophils (Lou 2007).

In order to analyse the physiological relevance of CD99 for leukocyte extravasation in vivo, we cloned mouse CD99, a gene with only 45% sequence homology to human CD99. Immunizing with a CD99-Fc fusion protein we generated polyclonal antibodies against mCD99. Affinity purified antibodies of this serum were able to inhibit the migration of antigen-stimulated T lymphocytes through monolayers of endothelial cells in vitro. Transmigration required CD99 on lymphocytes as well as on endothelial cells, as antibodies blocked transmigration on incubation with either of the two cell types (Bixel et al. 2004). In vivo, anti-CD99 antibodies inhibited the migration of radioactively labelled antigen-specific T cells into inflamed skin of the ear in a typical delayed type hypersensitivity (DTH) reaction. At the same time, these antibodies blocked ear swelling. Interestingly, homing of naïve T cells into lymph nodes was not inhibited by anti-CD99 antibodies.

Mouse as well as human CD99 is able to support homotypic cell adhesion, as transfected CHO cells aggregate in a CD99 dependent, Ca^{2+}-requiring manner. This is in agreement with the finding that CD99 is required on the leukocyte as well as on the endothelial side during the diapedesis process, suggesting a homophilic type interaction of CD99 during this process. However, we were not able to demonstrate directly that CD99 transfected cells bind to immobilized CD99-Fc or that beads coated with CD99-Fc would bind to CD99 transfected cells. Likewise, we could block neither leukocyte diapedesis or CD99-dependent aggregation of transfected CHO cells with monomeric F(ab) against mouse CD99 (Bixel et al. 2004). This argues against a homophilic interaction of CD99 as the basis for leukocyte diapedesis.

When mouse neutrophils were analysed, we found that anti-CD99 antibodies blocked diapedesis in vitro as well as neutrophil recruitment into inflamed peritoneum (Bixel et al. 2007). Interestingly and in contrast to lymphocytes, the antibodies blocked on the endothelial side only and not on the neutrophil side, although neutrophils expressed CD99. Using the TNF-α stimulated cremaster model we found that leukocyte extravasation was indeed blocked during diapedesis whereas leukocyte capturing to the blood vessel wall was unaffected (Bixel et al. 2007).

A distantly related protein of unknown function sharing 32% sequence homology with CD99 had been recently described as CD99L2 (Suh et al. 2003). CD99 as well as CD99L2 belong to none of the known gene families. They represent rather small type I membrane proteins with only one transmembrane region, a small intracellular part and an extracellular domain not much larger than 100 amino acids. Both proteins are highly O-glycosylated. We generated antibodies against mouse CD99L2 and showed that it is expressed on most leukocytes as well as on endothelial cells. Surprisingly, and in contrast to CD99, lymphocytes did not require CD99L2 for the diapedesis process, whereas anti-CD99L2 antibodies inhibited the migration of neutrophils into inflamed peritoneum (Bixel et al. 2007). As for anti-CD99 antibodies, anti-CD99L2 antibodies blocked the diapedesis process, but not leukocyte capturing in venules of TNF-α-stimulated cremaster muscle. Electron microscopy revealed that leukocytes did not get trapped at endothelial junctions between endothelial cells, but rather between the endothelial cells and the basement membrane (Bixel et al. 2007).

6 Concluding Remarks

The increasing number of endothelial membrane proteins demonstrated to be involved in diapedesis illustrates the complexity of this process. As was found many years ago for the capturing process, leukocyte diapedesis seems to be a multi-step process. Obviously, migration through the endothelial barrier and the barrier of the basement membrane are two different consecutive steps that require different molecules and mechanisms. In addition, migration through the endothelial cell layer already requires several different adhesion receptors fulfilling different tasks, such as: (1) transmission of leukocyte signals to the endothelial cells, which trigger the opening of endothelial cell contacts; (2) loosening of the endothelial cell contacts by molecules that regulate the adhesion molecules (e.g., VE-cadherin) that maintain the endothelium contact barrier; (3) support of the migration process by hetero- and homophilic adhesion molecules on endothelium and leukocytes. Elucidating the molecular machinery that controls and mediates this process will lead to the identification of novel pharmacological targets to control the process of inflammation.

References

Arrate MP, Rodriguez JM, Tran TM, Brock TA, Cunningham SA (2001) Cloning of human junctional adhesion molecule 3 (JAM-3) and its identification as the JAM-2 counter receptor. J Biol Chem 276:45826–45832

Aurrand-Lions M, Johnson-Leger C, Wong C, Du Pasquier L, Imhof BA (2001) Heterogeneity of endothelial junctions is reflected by differential expression and specific subcellular localization of the three JAM family members. Blood 98:3699–3707

Baumer S, Keller L, Holtmann A, Funke R, August B, Gamp A, Wolburg H, Wolburg-Buchholz K, Deutsch U, Vestweber D (2006) Vascular endothelial cell specific phospho-tyrosine phosphatase (VE-PTP) activity is required for blood vessel development. Blood 107:4754–4762

Bixel G, Kloep S, Butz S, Petri B, Engelhardt B, Vestweber D (2004) Mouse CD99 participates in T cell recruitment into inflamed skin. Blood 104:3205–3213

Bixel MG, Petri B, Khandoga AG, Khandoga A, Wolburg-Buchholz K, Wolburg H, Marz S, Krombach F, Vestweber D (2007) A CD99-related antigen on endothelial cells mediates neutrophil but not lymphocyte extravasation in vivo. Blood 109:5327–5336

Cera MR, Del Prete A, Vecchi A, Corada M, Martin-Padura I, Motoike T, Tonetti P, Bazzoni G, Vermi W, Gentili F, Bernasconi S, Sato TN, Mantovani A, Dejana E (2004) Increased DC trafficking to lymph nodes and contact hypersensitivity in junctional adhesion molecule-A-deficient mice. J Clin Invest 114:729–738

Chavakis T, Keiper T, Matz-Westphal R, Hersemeyer K, Sachs UJ, Nawroth PP, Preissner KT, Santoso S (2004a) The junctional adhesion molecule-C promotes neutrophil transendothelial migration in vitro and in vivo. J Biol Chem 279:55602–55608

Chavakis T, Preissner KT, Santoso S (2004b) Leukocyte trans-endothelial migration: JAMs add new pieces to the puzzle. Thromb Haemost 89:13–17

Corada M, Mariotti M, Thurston G, Smith K, Kunkel R, Brockhaus M, Lampugnani MG, Martin-Padura I, Stoppacciaro A, Ruco L, Mc Donald DM, Ward PA, Dejana E (1999) Vascular endothelial-cadherin is an important determinant of microvascular integrity in vivo. Proc Natl Acad Sci USA 96:9815–9820

Corada M, Chimenti S, Cera MR, Vinci M, Salio M, Fiordaliso F, De Angelis N, Villa A, Bossi M, Staszewsky LI, Vecchi A, Parazzoli D, Motoike T, Latini R, Dejana E (2005) Junctional adhesion molecule-A-deficient polymorphonuclear cells show reduced diapedesis in peritonitis and heart ischemia-reperfusion injury. Proc Natl Acad Sci USA 102:10634–10639

Dangerfield J, Larbi KY, Huang MT, Dewar A, Nourshargh S (2002) PECAM1 (CD31) Homophilic interaction up-regulates alpha 6 beta1 on transmigrated neutrophils In vivo and plays a functional role in the ability of alpha 6 integrins to mediate leukocyte migration through the perivascular basement membrane. J Exp Med 196:1201–1211

Daniel JM, Reynolds AB (1997) Tyrosine phosphorylation and cadherin/catenin function. Bioessays 19:883–891

Del Maschio A, De Luigi A, Martin-Padura I, Brockhaus M, Bartfai T, Fruscella P, Adorini L, Martino G, Furlan R, De Simoni MG, Dejana E (1999) Leukocyte recruitment in the cerebrospinal fluid of mice with experimental meningitis is inhibited by an antibody to junctional adhesion molecule (JAM). J Exp Med 190:1351–1356

Duncan GS, Andrew DP, Takimoto H, Kaufman SA, Yoshida H, Spellberg J, de la Pompa L, Elia A, Wakeham A, Karan-Tamir B, Muller WA, Senaldi G, Zukowski MM, Mak TW (1999) Genetic evidence for functional redundancy of platelet/endothelial cell adhesion molecule-1 (PECAM-1): CD31-deficient mice reveal PECAM-1-dependent and PECAM-1-independent functions. J Immunol 162:3022–3030

Ebnet K, Aurrand-Lions M, Kuhn A, Kiefer F, Butz S, Zander K, Brickwedde MK, Suzuki A, Imhof BA, Vestweber D (2003) The junctional adhesion molecule (JAM) family members JAM-2 and JAM-3 associate with the cell polarity protein PAR-3: a possible role for JAMs in endothelial cell polarity. J Cell Sci 116:3879–91

Esser S, Lampugnani MG, Corada M, Dejana E, Risau W (1998) Vascular endothelial growth factor induces VE-cadherin tyrosine phosphorylation in endothelial cells. J Cell Sci 111:1853–1865

Fachinger G, Deutsch U, Risau W (1999) Functional interaction of vascular endothelial-protein-tyrosine phosphatase with the angiopoietin receptor Tie-2. Oncogene 18:5948–5953

Gotsch U, Borges E, Bosse R, Böggemeyer E, Simon M, Mossmann H, Vestweber D (1997) VE-cadherin antibody accelerates neutrophil recruiment in vivo. J Cell Sci 110:583–588

Hirata K, Ishida T, Penta K, Rezaee M, Yang E, Wohlgemuth J, Quertermous T (2001) Cloning of an immunoglobulin family adhesion molecule selectively expressed by endothelial cells. J Biol Chem 276:16223–16231

Huang MT, Larbi KY, Scheiermann C, Woodfin A, Gerwin N, Haskard DO, Nourshargh S (2006) ICAM-2 mediates neutrophil transmigration in vivo: Evidence for stimulus-specificity and a role in PECAM-1-independent transmigration. Blood 107:4721–4727

Ishida T, Kundu RK, Yang E, Hirata K, Ho YD, Quertermous T (2003) Targeted disruption of endothelial cell-selective adhesion molecule inhibits angiogenic processes in vitro and in vivo. J Biol Chem 278:34598–604

Khandoga A, Kessler JS, Meissner H, Hanschen M, Corada M, Motoike T, Enders G, Dejana E, Krombach F (2005) Junctional adhesion molecule-A deficiency increases hepatic ischemia-reperfusion injury despite reduction of neutrophil transendothelial migration. Blood 106:725–733

Lampugnani MG, Resnati M, Raiteri M, Pigott R, Pisacane A, Houen G, Ruco LP, Dejana E (1992) A novel-endothelial specific membrane protein is a marker of cell-cell contacts. J Cell Biol 118:1511–1522

Lampugnani MG, Corada M, Andriopoulou P, Esser S, Risau W, Dejana E (1997) Cell confluence regulates tyrosine phosphorylation of adherens junction components in endothelial cells. J Cell Sci 110:2065–2077

Liao F, Huynh HK, Eiroa A, Greene T, Polizzi E, Muller WA (1995) Migration of monocytes across endothelium and passage through extracellular matrix involve separate molecular domains of PECAM-1. J Exp Med 182:1337–1343

Liao F, Ali J, Greene T, Muller WA (1997) Soluble domain 1 of platelet-endothelial cell adhesion molecule (PECAM) is sufficient to block transendothelial migration in vitro and in vivo. J Exp Med 185:1349–1357

Liu Y, Nusrat A, Schnell FJ, Reaves TA, Walsh S, Pochet M, Parkos CA (2000) Human junction adhesion molecule regulates tight junction resealing in epithelia. J Cell Sci 113:2363–2374

Lou O, Alcaide P, Luscinskas FW Muller WA (2007) CD99 is a key mediator of the transendothelial migration of neutrophils. J Immunol 178:1136–1143

Martin-Padura I, Lostaglio S, Schneemann M, Williams L, Romano M, Fruscella P, Panzeri C, Stoppacciaro A, Ruco L, Villa A Simmons D, Dejana E (1998) Junctional adhesion molecule, a novel member of the immunoglobulin superfamily that distributes at intercellular junctions and modulates monocyte transmigration. J Cell Biol 142:117–127

Muller WA, Weigl SA, Deng X, Phillips DM (1993) PECAM-1 is required for transendothelial migration of leukocytes. J Exp Med 178:449–460

Nasdala I, Wolburg-Buchholz K, Wolburg H, Kuhn A, Ebnet K, Brachtendorf G, Samulowitz U, Kuster B, Engelhardt B, Vestweber D, Butz S (2002) A transmembrane tight junction protein selectively expressed on endothelial cells and platelets. J Biol Chem 277:16294–16303

Nawroth R, Poell G, Ranft A, Samulowitz U, Fachinger G, Golding M, Shima DT, Deutsch U, Vestweber D (2002) VE-PTP and VE-cadherin ectodomains interact to facilitate regulation of phosphorylation and cell contacts. EMBO J 21:4885–4895

O'Brien CD, Lim P, Sun J, Albelda SM (2003) PECAM-1-dependent neutrophil transmigration is independent of monolayer PECAM-1 signaling or localization. Blood 101:2816–2825

Reiss Y, Hoch G, Deutsch U, Engelhardt B (1998) T cell interaction with ICAM-1-deficient endothelium in vitro: essential role for ICAM-1 and ICAM-2 in transendothelial migration of T cells. Eur J Immunol 28:3086–3099

Reymond N, Imbert AM, Devilard E, Fabre S, Chabannon C, Xerri L, Farnarier C, Cantoni C, Bottino C, Moretta A, Dubreuil P, Lopez M (2004) DNAM-1 and PVR regulate monocyte migration through endothelial junctions. J Exp Med 199:1331–1341

Schenkel AR, Mamdouh Z, Chen X, Liebman RM, Muller WA (2002) CD99 plays a major role in the migration of monocytes through endothelial junctions. Nat Immunol 3:143–150

Schenkel AR, Chew TW, Muller WA (2004) Platelet endothelial cell adhesion molecule deficiency or blockade significantly reduces leukocyte emigration in a majority of mouse strains. J Immunol 173:6403–6408

Shaw SK, Bamba PS, Perkins BN, Luscinskas FW (2001) Real-time imaging of vascular endothelial-cadherin during leukocyte transmigration across endothelium. J Immunol 167:2323–2330

Suh YH, Shin YK, Kook MC, Oh KI, Park WS, Kim SH, Lee IS, Park HJ, Huh TL, Park SH (2003) Cloning, genomic organization, alternative transcripts and expression analysis of CD99L2, a novel paralog of human CD99, and identification of evolutionary conserved motifs. Gene 307:63–76

Ukropec JA, Hollinger M K, Salva S M, Woolkalis MJ (2000) SHP2 association with VE-cadherin complexes in human endothelial cells is regulated by thrombin. J Biol Chem 275:5983–5986

Wakelin MW, Sanz MJ, Dewar A, Albelda SM, Larkin SW, Boughton Smith N, Williams TJ, Nourshargh S (1996) An anti-platelet-endothelial cell adhesion molecule-1 antibody inhibits leukocyte extravasation from mesenteric microvessels in vivo by blocking the passage through the basement membrane. J Exp Med 184:229–239

Wegmann F, Ebnet K, Du Pasquier L, Vestweber D, Butz S (2004) Endothelial adhesion molecule ESAM binds directly to the multidomain adaptor MAGI-1 and recruits it to cell contacts. Exp Cell Res 300:121–33

Wegmann F, Petri J, Khandoga AG, Moser C, Khandoga A, Volkery S, Li H, Nasdala I, Brandau O, Fässler R, Butz S, Krombach F, Vestweber D (2006) ESAM supports neutrophil extravasation, activation of Rho and VEGF-induced vascular permeability. J Exp Med 203:1671–1677

Woodfin A, Reichel CA, Khandoga A, Corada M, Voisin MB, Scheiermann C, Haskard DO, Dejana E, Krombach F, Nourshargh S (2007) JAM-A mediates neutrophil transmigration in a stimulus-specific manner in vivo: evidence for sequential roles for JAM-A and PECAM-1 in neutrophil transmigration. Blood May 15 [Epub ahead of print]

Yamada S, Pokutta S, Drees F, Weis WI, Nelson WJ (2005) Deconstructing the cadherin-catenin-actin complex. Cell 123:889–901

Zen K, Babbin BA, Liu Y, Whelan JB, Nusrat A, Parkos CA (2004) JAM-C is a component of desmosomes and a ligand for CD11b/CD18-mediated neutrophil transepithelial migration. Mol Biol Cell 15:3926–3937

Ernst Schering Foundation Symposium Proceedings, Vol. 3, pp. 169–185
DOI 10.1007/2789_2007_064

Published Online: 18 December 2007

Fragment-Based Drug Discovery Using Rational Design

H. Jhoti(✉)

Astex Therapeutics Ltd., 436 Science Park, Milton Rd, CB40QA Cambridge, UK
email: *H.Jhoti@Astex-therapeutics.com*

Abstract. Fragment-based drug discovery (FBDD) is established as an alternative approach to high-throughput screening for generating novel small molecule drug candidates. In FBDD, relatively small libraries of low molecular weight compounds (or fragments) are screened using sensitive biophysical techniques to detect their binding to the target protein. A lower absolute affinity of binding is expected from fragments, compared to much higher molecular weight hits detected by high-throughput screening, due to their reduced size and complexity. Through the use of iterative cycles of medicinal chemistry, ideally guided by three-dimensional structural data, it is often then relatively straightforward to optimize these weak binding fragment hits into potent and selective lead compounds. As with most other lead discovery methods there are two key components of FBDD; the detection technology and the compound library. In this review I outline the two main approaches used for detecting the binding of low affinity fragments and also some of the key principles that are used to generate a fragment library. In addition, I describe an example of how FBDD has led to the generation of a drug candidate that is now being tested in clinical trials for the treatment of cancer.

1 Introduction

In the last 10 years we have witnessed the emergence of fragment-based drug discovery (FBDD), a new way of generating small molecule drug candidates, within the drug industry. The approach provides an alternative to more conventional methods such as high-throughput screening (HTS) and, although still a niche activity in pharmaceutical and biotechnology companies, it is rapidly increasing in popularity. This is reflected in the growing number of reviews on the subject (Rees et al. 2004; Erlanson et al. 2004; Hajduk 2006; Hajduk and Greer 2007). Although FBDD is unlikely to supplant HTS totally it does provide an alternative to screening targets, in particular, those that have their three-dimensional structures resolved.

The fundamental thesis of FBDD proposes that very small and simple chemical entities (MW 100–200), or fragments, are used to probe the surface of a target protein in order to identify regions that have the highest potential for interactions, that is, 'hot spots'. Given their very limited functional groups these fragments are expected to bind with low-affinity, if at all. This should, however, not be confused with sub-optimal binding as in order to overcome the not insignificant entropic barriers the fragment must display 'high ligand efficiency' in order to bind at all (Murray and Verdonk 2002). The concept of ligand efficiency, a measure derived from dividing the affinity of a small molecule by its size, is a very instructive way to normalize binding of high and low potency compounds with respect to size (Kuntz 1999; Hopkins 2004). It has now been established that many low-affinity (but high ligand efficiency) fragment hits can be developed into novel, potent lead compounds using fragment hit-to-lead chemistry (Hajduk and Greer 2007). This process is made significantly more efficient if the three-dimensional structure of the fragment bound to the target protein is available. In this context, iterative cycles of protein structure-guided medicinal chemistry have been highly successful in exploiting fragment hits (Gill et al. 2005; Carr and Jhoti 2002).

In March 2007 there were seven compounds publicly disclosed as undergoing or approved for clinical trials, whose origins lie in fragment-based screening: ABT-263 (Bcl-2 inhibitor) and ABT-518 (MMP inhibitor) from Abbott Laboratories, AT7519 (CDK2 inhibitor) and AT9283 (Aurora kinase inhibitor) from Astex Therapeutics, PLX204 (PPAR inhibitor) and PLX4032 (BRAF inhibitor) from Plexxicon and LP261 (tubulin inhibitor) from Locus. Three more IND (investigational new drug) applications are expected within the next 12 months: SGX523 (c-Met kinase) from SGX, SNS314 (Aurora kinases) from Sunesis and AT13387 (HSP90) from Astex Therapeutics.

One reason for this high level of productivity has been the ability of FBDD to generate multiple lead series against a given target. Although fragment libraries are small (typically fewer than 5×10^3 compounds) compared to their HTS counterparts (typically more than 1×10^6), fragments are intrinsically more promiscuous as they lack redundant complexity which in most cases will reduce binding to a protein active site often due to steric hinderance (Hann et al. 2001). Therefore, more chemical space can be explored by fragment screening due to their lower molecular complexity. Thus both hit-rates and the number of different chemotypes which are detected as hits, and which may therefore be developed into lead series, are usually greater for FBDD than for HTS. The greater number of lead series increases the chances of finding a lead compound which is suitable for pre-clinical development and furthermore, the lead compound should have a lower molecular weight (compared to one derived from HTS) which may impart better drug-like properties (Rees et al. 2004).

Biophysical techniques such as X-ray crystallography and NMR have been most extensively used for fragment screening as not only are they capable of detecting very low-affinity binding interactions (unlike conventional bioassays) but they can also provide structural information to guide the fragment optimization process (Jahnke and Erlanson 2006; Jhoti and Leach 2007). NMR was used in some of the earliest fragment screening experiments by the Abbott group who pioneered the SAR-by-NMR approach (Shuker et al. 1996). More recently, X-ray crystallography has become popular as it is able to elucidate the exact binding mode of fragments and thus provide medicinal chemistry design strategies to evolve a fragment into a potent, drug-sized selective lead compound (Nienaber et al. 2000; Hartshorn et al. 2005). As previously mentioned, despite the low affinity of initial fragment hits, useful fragments typically exhibit a high ligand efficiency, that is, a high value for the average free energy of binding per heavy atom (i.e. excluding hydrogens) (Hopkins et al. 2004). It is critical that during the development into the lead compound, this high ligand efficiency is maintained.

Although fragment-based screening typically seeks to identify ligands bound to a predefined site on a protein (Fig. 1a) it is often possible to identify alternative ligand binding sites and/or observe simultaneous fragment binding at spatially distinct sites on the protein (Fig. 1b). Fragments that are observed to bind in relatively close proximity to one another provide data on the ways in which different functionalities can be combined (linked) within a single, ideally higher potency, ligand (Fig. 1c). Analysis of the binding of individual fragments can help to define sensible growth vectors off a template molecule (Fig. 1d). Such analyses can be of great benefit if fragment binding induces conformational changes within a protein, revealing previously inaccessible, or structurally modified, binding pockets (Rees et al. 2004).

2 Fragment Detection

Fragment screening platforms today are becoming more sophisticated as the different biophysical techniques (e.g. X-ray, NMR,) are being integrated with the result of improved hit rates. For example, the Astex group have integrated NMR, high-throughput X-ray crystallography, calorimetry together with virtual screening into a fragment-based discovery platform—Pyramid. Central to Pyramid is AutoSolve, a web-based software application linked to an in-house developed LIMS system and Oracle database that performs automated processing and analysis of all protein-ligand X-ray data (Mooij et al. 2006). Other companies such as Structural GenomiX (SGX) have also developed platforms for lead compound identification using high-throughput protein structure determination: SGX's *FAST*™ (Fragments of Active Structures Technology) enables the rapid identification of novel, potent, and

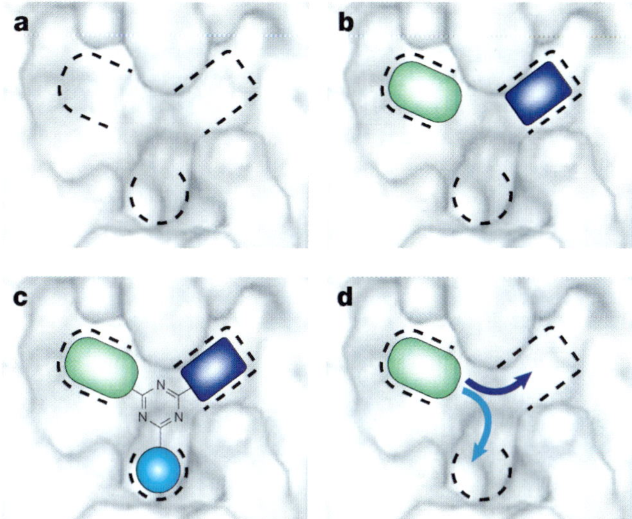

Fig. 1. **a**. A region of interest is defined on the protein surface; **b**: two fragments bound to distinct sites (pockets) within the region of interest; **c**: synthesis of a ligand that combines the fragments identified in b, but also accesses a previously unexplored pocket on the protein surface; **d**: a fragment (template molecule) to a single pocket on the protein suggests potential growth vectors. From Jhoti 2003

selective small-molecule inhibitors of drug targets. Card et al. (2005) at Plexxikon have described their process Scaffold-Based Drug Discovery™ for the design of molecular scaffolds and ligands, based on X-ray analysis of co-crystals of protein and ligand. Despite the diversity of biophysical approaches now being used for fragment screening, the most commonly used methods remain X-ray crystallography and Ligand directed NMR, therefore the remainder of this section will focus on these techniques.

2.1 X-Ray Crystallography

Some of the first experiments in which X-ray crystallography was used as a fragment screening tool were reported by Verlinde et al. (1997) who exposed crystals of trypanosomal triosephosphate isomerase to cocktails (mixtures) of compounds in the search for inhibitors. Soon after, Nienaber et al. (2000) at Abbott Laboratories described their CrystaLEAD™ process for X-ray-based screening of shape diverse fragment sets. As with many screening techniques, X-ray crystallographic fragment screening is most efficiently carried out in a cocktail format. Typically, a complete library (500–1000 fragments) is partitioned into a set of multi-fragment cocktails, each containing an equal number of components (typically 4–10). Individual crystals of the target protein are then soaked in these cocktails and X-ray diffraction data are collected from the soaked crystals (Jhoti 2003). The final concentration of the compounds in the cocktails is relatively high, in the region of 25–100 mM, as it is expected that affinities for fragments will lie in the high micromolar to low millimolar range (Hartshorn et al. 2005; Berdini et al. 2007). Alternatively, fragments may be co-crystallized with the target protein as described by Card et al. (2005); however, this approach is usually limited to individual fragments and is not applicable to cocktails. The resulting electron density maps reveal whether one or more of the fragments have bound to sites on the protein. In the event of multiple fragments competing for the same site it may be necessary to deconvolute the cocktail by partitioning it into smaller cocktails and/or individual fragment soaks. Identification of the exact binding mode of a ligand can also be made easier by using ligand libraries in which the ligands are decorated with electron-dense substituent groups, such as halogens (exemplified by SGX's FAST™).

Screening by X-ray crystallography requires the production of large numbers of diffraction quality crystals, either in the absence of compound for soaking experiments, or in the presence of compound for co-crystallization. It is therefore necessary to be able to generate crystals of a similar size and quality on a large scale (several hundreds) reproducibly. This requires optimization of the processes from protein production to crystallization. Progress has been made in protein production, partly due to the efforts of structural genomics initiatives and partly due to the introduction of robotics that has enabled parallelization of expression trials, increasing the chances of producing soluble active proteins. Forstner et al. (2007) review the recent improvements in expression systems and methodology of cloning and expression. The use of affinity tags has been invaluable in enabling rapid purification of proteins and most structural genomics initiatives incorporate some form of tag in their expression vectors.

Another area of progress is in the techniques for crystallization. Automated systems for crystallization screening are now able to screen large numbers of crystallization conditions using a very small amount of protein (Stevens 2000). Once initial crystallization conditions have been obtained, robots can also help to generate consistent crystals for routine screening. Crystals used for fragment screening are necessarily exposed to high concentrations of a wide variety of compounds and an important part of the screening process is the optimization of the protocol by which the crystals are handled during soaking and freezing. This includes stabilization of the crystal by varying factors which include pH, buffer and precipitant concentration as well as cryoprotectant (Blundell et al. 2002).

The use of X-ray crystallography as a screening tool requires that systems for rapid, efficient collection of X-ray diffraction data must be in place, as collection of diffraction data for many crystals could conceivably be a slow step in the screening cascade (Muchmore et al. 2000). Access to bright synchrotron sources such as the macromolecular crystallography beamlines at the ESRF in Grenoble enables a high rate of data collection, and increased automation has increased the throughput even further. An automated data collection and processing system has been installed on beamlines at the ESRF and has been interlinked with structural databases (Beteva et al. 2006). Much of this automation is now also available for laboratory X-ray sources which, together with improved X-ray intensities, allow high throughput data collection to be performed in-house.

A key step in high throughput protein–ligand crystallography is the automatic placement of the ligand into the experimentally observed electron density, a task which until recently was carried out mostly manually. There are two drawbacks to the manual interpretation of ligand density. Firstly, the placement can be biased by the crystallographer's knowledge of the target and ligand. Secondly, and critical to crystallographic screening approaches, it is slow. Recently, there have been reports in the literature on the automatic placement of ligands into electron density, including the X-Ligand approach (Oldfield 2001) and the ligand building routines in ARP/wARP (Evrard et al. 2007). Mooij and colleagues (2006) have also developed the AutoSolve software to address this issue. The ligand-fitting methodology in AutoSolve is developed around the protein–ligand docking program GOLD (Verdonk et al. 2003). Another feature of AutoSolve is that it tightly integrates the ligand-placement routines with X-ray data processing, molecular replacement, as well as water placement and protein/ligand refinement cycles. All data (structure factor files, molecular replacement models, binding site definitions, ligand SMILES) are stored in Oracle databases, linked to web-based interfaces to run AutoSolve and to visualize the resulting structures. Linking all these aspects of solving the X-ray structure of a protein–ligand complex in AutoSolve has meant that datasets can often be taken from intensity scaling to a refined protein–ligand complex with the minimum of user intervention.

2.2 Ligand-Detected NMR Experiments

While some of the first applications of fragment screening used two-dimensional NMR (also referred to as 'protein detected') methods in which the spectra of the protein are examined to detect evidence of ligand binding (i.e., SAR-by-NMR), methods that focus on perturbations of the ligand spectra have become more popular (Jahnke and Erlanson 2006; Jhoti and Leach 2007). Ligand-detected NMR experiments have the advantage that no restriction is placed on the size of the target protein and that there is no necessary requirement for isotope-labelling (Lepre and Moore 2007). Binding of the fragment to the target is detected via changes in NMR parameters of the fragment. The parameters that are most strongly affected by protein binding include relaxation (T1, T2, T1ρ) and cross relaxation (NOE, water-LOGSY, STD) effects. Changes in the average values of these parameters can be detected when less than 0.5% of the ligand is complexed with protein, thus they provide a sensitive measure of binding. Ligand-detected NMR is also able to detect fragments whose limited solubility may be a constraint when using X-ray crystallography as the required concentration of the fragment is an order of magnitude lower (rather than higher) than its Kd. For example, a fragment with a Kd of 1 mM will typically need to be soaked at a concentration of around 10 mM into a crystal to ensure good occupancy but by NMR it needs to be screened at only 0.1 mM.

One original drawback of ligand-detected NMR screening was that, in contrast to protein-detected NMR methods, no information on the binding mode of the ligand could be obtained. However, recent developments have exploited competitive ligand binding to add information on the binding mode and affinity of a fragment and also to reduce the amount of protein that may be required. Further detail on the position and orientation of the bound fragment can also be obtained from saturation-transfer difference (STD) NMR experiments (McCoy et al. 2005). Saturation transfer between protons of the protein and a bound ligand has been used to map the parts of the ligand that are in direct contact with the protein (Mayer and Meyer 2001). Similarly, STD experiments using protein that has been deuterated everywhere, except for specific amino acid types, can be used to determine the residue types in the binding site and their orientation with respect to the ligand (Hadjuk et al. 2004). Of particular interest for drug design is the use of STD to determine the relative binding orientations of competitive fragment hits—the INPHARMA method (Sanchez-Pedregal et al. 2005). This approach does not require protein deuteration but instead relies on the indirect transfer of magnetization between specific protons of two competitive ligands, mediated by a small number of protons of the target. In cases where fragment-based screening identifies many structurally diverse hits, INPHARMA should allow the rapid determination of their relative orientations, from which common epitopes may be inferred and selected series prioritized for detailed structural studies.

3 Fragment Libraries

In general, fragment libraries typically have compounds that have molecular weights of 100–250 Da and are relatively simple with few functional groups, making them chemically suitable for rapid synthetic optimization. Another key advantage of their small size is their potential to sample more chemical space and have a higher probably of binding due to the Hann model of molecular complexity (Hann et al. 2001). This feature of more efficient chemical sampling results in the size of the typical fragment library being significantly smaller (500–5000 fragments) than compound libraries used in HTS which can reach over 1 million compounds. Given the requirement that the fragments need to be screened at relatively high concentrations, maybe even up to 50 mM, the solubility of the fragments need to be carefully examined before inclusion into the library. In addition to solubility there are several other physico-chemical properties that a 'ligand-efficient' fragment needs to exhibit and many of these have been outlined in a 'Rule-of-three' filter that was defined by the Astex group and is now widely used (Congreve et al. 2003). As outlined above, there are different biophysical techniques which can be used for fragment screening, and each technique is likely to apply its own constraints, such as different solubility requirements, on the features of the fragment library. Therefore, separate libraries are often generated that are more suitable for screening with, for example, X-ray or NMR to exploit the strengths of the detection techniques.

Some of the general approaches that are widely used to construct fragment libraries include: virtual screening of the target protein, ligand-based analysis and sampling known drug diversity (Berdini et al. 2007). The first two exploit knowledge of the target or ligand and can be used to generate fragment libraries that are useful in screening protein families such as kinases, phosphatases or proteases. Virtual screening methods tend to involve docking of fragments into the protein structure and can identify fragments that should be included in the screening set (Verdonk et al. 2003). Given the significant challenges of working with fragments, there should be less emphasis on trying to predict the correct orientation in the active sites as compared with docking larger compounds. Ligand-based analysis can consist of analysing (or even fragmenting) known inhibitors and substrates to identify fragments that have features of the chemotype that is recognized by the protein. Unlike the first two, the third approach is not 'knowledge-based' but attempts to explore known drug space by analysing bioactive compounds and identifying fragments that represent some of the chemotypes observed in these compounds (Hartshorn et al. 2005).

3.1 Fragment to Clinic

As mentioned earlier there are at least seven compounds originating from a fragment screening approach that are approved for or already in clinical testing. One of these compounds is AT7519 (from Astex) which is a potent Cyclin-dependent kinase 2 (CDK2) inhibitor and is currently in multiple phase 1 trials for patients with solid tumours.

CDK2 is a representative of a family of Ser/Thr protein kinases involved in the cell cycle. CDK inhibitors are thought to have potential in anti-cancer therapy (Fischer and Lane 2000). The architecture of the adenosine 5′-triphosphate (ATP) binding site of protein kinases is structurally well understood and CDK2 behaves as a classic case (Gill 2004). Using the Pyramid fragment discovery approach developed at Astex, crystals of CDK2 were soaked with the Focused Kinase Set, which provided 29 hits out of 212 fragments (13.7% hit rate) (Berdini et al 2007). As observed with other fragment screens, these fragment hits exhibit affinities (IC_{50}) in the millimolar to low micromolar range. The relatively large number of hits is also a general feature of fragment screening, consistent with the premise that efficient sampling of chemical space is possible even using only a few hundred fragments.

Two examples of the hits identified in the CDK2 fragment screen are shown in Fig. 2. These fragments make good H-bonding interactions with the 'hinge region' of CDK2 and are clearly defined in the electron density. Both bind with an orientation that would allow subsequent optimization to generate a potent lead compound. A fragment-to-lead optimization is exemplified using the indazole fragment in Fig. 3. After analysing the binding mode of the fragment hit and also the local environment of the enzyme pocket small functional groups are attached to the indazole fragment using structure-guided medicinal chemistry. Typically, between 5 and 10 compounds are synthesized in each iterative cycle in a fragment-to-lead optimization process in an attempt to improve potency or other aspects of binding. Crystal structures are usually obtained of most of these compounds and one is shown in the centre panel of Fig. 3. After additional cycles of structure-guided chemistry, the resulting compound often has nanomolar affinity such as the final compound in Fig. 3 (Berdini et al. 2007).

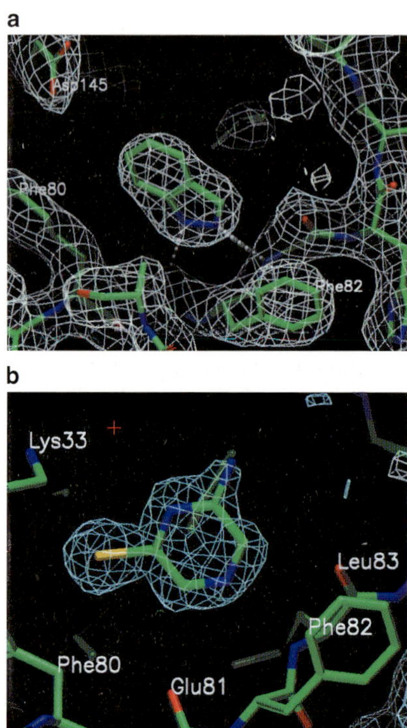

Fig. 2. a Fragment from the Focused Kinase Screen bound at the active site of CDK2 ($2F_o$–F_c map at 1σ); **b** CDK2 active site showing fragment identified by virtual screening (F_o–F_c map at 3σ)

Several features of this fragment-to-lead optimization process should be emphasized. Firstly, as shown in the figure the original binding orientation of the fragment is conserved in the final compound. This is consistent with the premise that the original fragment hit has identified a strong 'anchor' interaction in the protein which is also reflected in a good ligand efficiency measure. Although at Astex we occasionally observe some movement of the original fragment hit, in most cases the original fragment binding mode is conserved. However, there have been reports from other groups that the original fragment binding mode

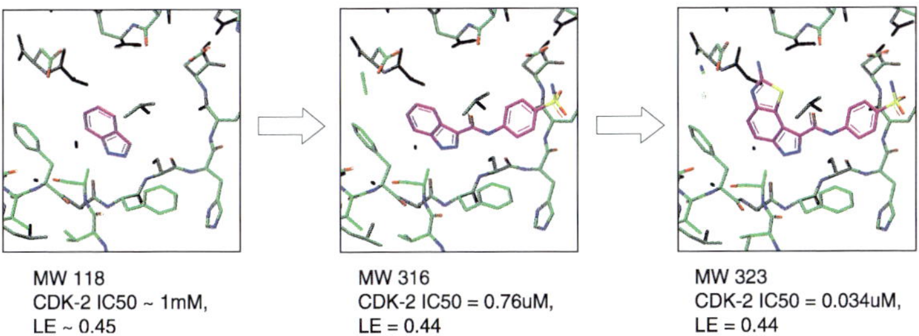

Fig. 3. Fragment-to-lead optimization process showing 'evolution' of weak fragment hit to potent lead compound in CDK2. Note how the ligand efficiency is maintained during the process of iterative structure-guided chemistry

can change more often and even dramatically. This maybe explained by (1) the fragments chosen are not exhibiting optimal binding (i.e., good ligand efficiency) or (2) the systems being screened, such as protein–protein interactions, obey different rules than the more rigid enzymes that we have explored (Oltersdorf et al. 2005). Another general feature that the indazole optimization process exemplifies is that the ligand efficiency is maintained while the fragment is 'grown'. This means that atoms are added only if they generate sufficient improvement in potency, which allows the overall size of the compound to be kept as small as possible, a key element in generating good drug-like molecules. It was a similar optimization process using another CDK2 fragment hit that resulted in AT7519 which is now in phase 1 clinical trials. The molecular weight of AT7519 is <400 Da, which is significantly lower than most other kinase inhibitors, and it has an LE measure of 0.4. This is consistent with the goal of using FBDD to generate highly efficient drug candidates with optimal properties which may result in new medicines.

4 Conclusion

FBDD is now established as one of the most exciting new approaches for the generation of novel small molecule drug candidates. This is reflected by the fact that most pharmaceutical companies are now active in this area. The technological advances seen in the last 10 years in structural biology have allowed high-throughput methods for X-ray crystallography and NMR to be applied in fragment screening. Integrating several of these biophysical techniques into a platform for detecting fragment binding has improved the efficiency of generating novel lead compounds. It will be in the next 5–10 years that we will learn whether this approach helps to improve the productivity of drug discovery—the ultimate goal of any new lead discovery technology.

Acknowledgements. I wish to thank all the staff at Astex Therapeutics for allowing me to present their work. In particular, I would thank Dr. Glyn Williams for guidance on the NMR techniques and Dr. Ian Tickle for X-ray crystallography.

References

Berdini V, O'Reilly M, Congreve MS, Tickle IJ (2007) Fragment-based screening by X-ray crystallography. Jhoti H, Leach AR (eds) Structure-based drug discovery. Springer, Berlin Heidelberg New York, pp 99–123

Beteva A, Cipriani F, Cusack S, Delageniere S, Gabadinho J, Gordon EJ, Guijarro M, Hall DR, Larsen S, Launer L, Lavault CB, Leonard GA, Mairs T, McCarthy A, McCarthy J, Meyer J, Mitchell E, Monaco S, Nurizzo D, Pernot P, Pieritz R, Ravelli RG, Rey V, Shepard W, Spruce D, Stuart DI, Svensson O, Theveneau P, Thibault X, Turkenburg J, Walsh M, McSweeney SM (2006) High-throughput sample handling and data collection at synchrotrons: embedding the ESRF into the high-throughput gene-to-structure pipeline. Acta Cryst D Biol Crystallogr 62:1162–1169

Blundell TL, Jhoti H, Abell C (2002) High-throughput crystallography for lead discovery in drug design. Nat Rev Drug Disc 1:45–54

Card GL, Blasdel L, England BP, Zhang C, Suzuki Y, Gillette S, Fong D, Ibrahim PN, Artis DR, Bollag G, Milburn MV, Kim SH, Schlessinger J, Zhang KY (2005) A family of phosphodiesterase inhibitors discovered by cocrystallography and scaffold-based drug design. Nat Biotechnol 23:201–207

Carr R, Jhoti H (2002) Structure-based screening of low-affinity compounds. Drug Discov Today 7:522–527

Congreve M, Carr R, Murray C, Jhoti H (2003) A 'rule of three'™ for fragment-based lead discovery? Drug Discov Today 8:876–877

Erlanson DA, McDowell RS, O'Brien T (2004a) Fragment-based drug discovery. J Med Chem 47:3463–3482

Evrard GX, Langer GG, Perrakis A, Lamzin VS (2007) Assessment of automatic ligand building in ARP/wARP. Acta Cryst D Biol Crystallogr 63:108–117

Fischer PM, Lane DP (2000) Inhibitors of cyclin-dependent kinases as anticancer therapeutics. Curr Med Chem 7:1213–1245.

Forstner M, Leder L, Mayr LM (2007) Optimization of protein expression systems for modern drug discovery. Expert Rev Proteomics 4:67–78

Gill AL (2004) New lead generation strategies for protein kinase inhibitors – fragment-based screening approaches, Mini-reviews Med Chem 4:301–311

Gill A, Frederickson M, Cleasby A, Woodhead SJ, Carr MG, Woodhead AJ, Walker MT, Congreve MS, Devine LA, Tisi D, Magor LCA, Davis DJ, Curry J, Anthony R, Padova A, Murray CW, Carr RAE, Jhoti H (2005) Identification of novel p38α MAP kinase inhibitors using fragment-based lead generation. J Med Chem 48:414–426

Hajduk PJ, Mack JC, Olejniczak ET, Park C, Dandliker PJ, Beutel BA (2004) SOS-NMR: a saturation transfer NMR-based method for determining the structures of protein-ligand complexes. J Am Chem Soc 126:2390–2398

Hajduk PJ (2006) Fragment-based drug design: how big is too big? J Med Chem 49:6972–6976

Hajduk PJ, Greer J (2007) A decade of fragment-based drug design: strategic advances and lessons learned. Nat Rev Drug Discov 6:211–219

Hann MM, Leach AR, Harper G (2001) Molecular complexity and its impact on the probability of finding leads for drug discovery. J Chem Inf Comp Sci 41:856–864

Hartshorn MJ, Murray CW, Cleasby A, Frederickson M, Tickle IJ, Jhoti H (2005) Fragment-based lead discovery using X-ray crystallography. J Med Chem 48:403–413

Hopkins AL, Groom CR, Alex A (2004) Ligand efficiency: a useful metric for lead selection. Drug Discovery Today 9:430–431

Jahnke W, Erlanson DA (2006) Fragment-based approaches in drug discovery. Wiley-VCH

Jhoti H (2003) High-throughput X-ray techniques and drug discovery. In: Waldman H, Koppitz M (eds) Small molecule-protein interactions. (Ernst-Schering Research Foundation) Springer, Berlin Heidelberg New York, pp 43–58

Jhoti H, Leach AR (2007) Structure-based drug discovery. Springer, Berlin Heidelberg New York

Kuntz ID, Chen K, Sharp KA, Kollman PA (1999) The maximal affinity of ligands. Proc Natl Acad Sci USA 96:9997–10002

Lepre CA, Moore JM (2007) Fragment-based NMR Screening in lead discovery. In: Jhoti H, Leach AR (eds) Structure-based drug discovery. Springer, Berlin Heidelberg New York, pp 73–98

Mayer M, Meyer B (2001) Group epitope mapping by saturation transfer difference NMR to identify segments of a ligand in direct contact with a protein receptor. J Am Chem Soc 123:6108–6117

McCoy MA, Senior MM, Wyss DF (2005) Screening of protein kinases by ATP-STD NMR spectroscopy. J Am Chem Soc 127:7978–7979

Mooij WT, Hartshorn MJ, Tickle IJ, Sharff AJ, Verdonk ML, Jhoti H (2006) Automated protein-ligand crystallography for structure-based drug design. Chem Med Chem 1:827–838

Muchmore SW, Olson J, Jones R, Pan J, Blum M, Greer J, Merrick SM, Magdalinos P, Nienaber VL (2000) Automated crystal mounting and data collection for protein crystallography. Structure 8:R243–R246

Murray CW, Verdonk ML (2002) The consequences of translational and rotational entropy lost by small molecules on binding to proteins. J Comp-Aided Mol Design 16:741–753

Nienaber VL, Richardson PL, Klighofer V, Bouska JJ, Giranda VL, Greer J (2000) Discovering novel ligands for macromolecules using X-ray crystallographic screening. Nat Biotechnol 18:1105–1108

Oltersdorf T, Elmore SW, Shoemaker AR, Armstrong RC, Augeri DJ, Belli BA, Bruncko M, Deckwerth TL, Dinges J, Hajduk PJ, Joseph MK, Kitada S, Korsmeyer SJ, Kunzer AR, Letai A, Li C, Mitten MJ, Nettesheim DG, Ng S, Nimmer PM, O'Connor JM, Oleksijew A, Petros AM, Reed JC, Shen W, Tahir SK, Thompson CB, Tomaselli KJ, Wang B, Wendt MD, Zhang H, Fesik SW, Rosenberg SH (2005) An inhibitor of Bcl-2 family proteins induces regression of solid tumours. Nature 435:677–681

Oldfield TJ (2001) X-LIGAND: an application for the automated addition of flexible ligands into electron density. Acta Cryst D Biol Crystallogr 57:696–705

Rees DC, Congreve M, Murray CW, Carr R (2004) Fragment-based lead discovery. Nat Rev Drug Discov 3:660–672

Sanchez-Pedregal VM, Reese M, Meiler J, Blommers MJ, Griesinger C, Carlomagno T (2005) The INPHARMA method: protein-mediated interligand NOEs for pharmacophore mapping. Angew Chem Int Ed Engl 44:4172–4175

Shuker SB, Hajduk PJ, Meadows RP, Fesik SW (1996) Discovering high-affinity ligands for proteins: SAR by NMR. Science 274:1531–1534

Stevens RC (2000) High-throughput protein crystallization, Curr Opin Struct Biol 10:558–563

Verdonk ML, Cole JC, Hartshorn MJ, Murray CW, Taylor RD (2003) Improved protein-ligand docking using GOLD. Proteins 52:609–623

Verlinde CLMJ, Kim H, Bernstein BE, Mande SC, Hol WGJ (1997) in Structure-based drug design. Edited by Veerapandian P. Marcel Dekker, New York, pp. 365–394

Ernst Schering Foundation Symposium Proceedings

Editors: Günter Stock
Monika Lessl

Printing: Krips bv, Meppel, The Netherlands
Binding: Stürtz, Würzburg, Germany